TEACHING THROUGH EXAMPLE-BASED PROBLEM SOLVING (TEPS) IN EARLY ALGEBRA

This elementary mathematics resource outlines a pedagogical approach known as Teaching Through Example-Based Problem Solving (TEPS) and presents pre-service teachers with an effective way to instruct early algebra. Using contrasting cases to illustrate inverse relations and the basic properties of operations, it breaks down an often controversial "gatekeeper" for students' mathematical learning and strengthens their algebraic thinking.

Conventional algebraic teaching often relies on a "show-and-tell" approach or focuses on surface-level questions, overlooking opportunities for student cognitive engagement and deep learning. Yet, TEPS integrates real-world contexts with concreteness fading and employs deep questioning to promote students' engagement and understanding. Each case in this illuminating book contains two lesson scripts or sub-cases: one illustrating the TEPS approach and the other demonstrating a more "traditional" approach, offering teachers opportunities to compare and optimize their own teaching. Ideal for use in pre-service elementary math methods courses, and for teacher professional development as a supplement to any elementary math textbook, this book and the TEPS approach empower teachers to meaningfully engage students in an often daunting subject area.

Meixia Ding is a Professor of Mathematics Education at Temple University, USA.

TEACHING THROUGH EXAMPLE-BASED PROBLEM SOLVING (TEPS) IN EARLY ALGEBRA

Contrasting Cases in Elementary Math

Meixia Ding

Routledge
Taylor & Francis Group
NEW YORK AND LONDON

Designed cover image: Getty Images

First published 2026
by Routledge
605 Third Avenue, New York, NY 10158

and by Routledge
4 Park Square, Milton Park, Abingdon, Oxon, OX14 4RN

Routledge is an imprint of the Taylor & Francis Group, an informa business

ISBN: 978-1-032-95747-0 (hbk)
ISBN: 978-1-032-95746-3 (pbk)
ISBN: 978-1-003-58633-3 (ebk)

DOI: 10.4324/9781003586333

Typeset in Times New Roman
by KnowledgeWorks Global Ltd.

To My Beloved Family: Xiaobao, Mohen, and Owen
Each step of my journey has been inseparable from your unwavering love and support.

CONTENTS

Acknowledgements ix

1 Introduction 1

1.1 Background 1
1.2 Goal of the Book 2
1.3 Why Contrasting Cases? 2
1.4 Why Early Algebra? 3
1.5 Pilot Work 3
1.6 Structure of the Book 4

2 Analytical Frameworks: TEPS and ICAP 6

2.1 TEPS: Teaching through Example-based Problem Solving 6
2.2 ICAP: Interactive, Constructive, Active, and Passive 7

3 Teaching Additive Inverses: Contrasting Cases 1–2 10

3.1 Teaching Additive Inverses with Part-whole Model 11
3.2 Teaching Additive Inverses with Comparison Model 23
3.3 Summary 36

4 Teaching Multiplicative Inverses: Contrasting Cases 3–5 38

4.1 Teaching Multiplicative Inverses with Equal-groups Model 38
4.2 Teaching Multiplicative Inverses with Array Model 50
4.3 Teaching Multiplicative Inverses with Comparison Model 63
4.4 Summary 77

5 Teaching Basic Properties of Addition: Contrasting Cases 6–7 79

5.1 Teaching Commutative Property of Addition 79
5.2 Teaching Associative Property of Addition 90
5.3 Summary 103

6 Teaching Basic Properties of Multiplication: Contrasting Cases 8–10 105

6.1 Teaching Commutative Property of Multiplication 105
6.2 Teaching Associative Property of Multiplication 116
6.3 Teaching Distributive Property of Multiplication Over Addition 129
6.4 Summary 142

7 Conclusion 144

7.1 Purpose of This Book 144
7.2 Revisiting the TEPS Approach 145
7.3 Revisiting Contrasting Cases and ICAP 147
7.4 Looking Back and Moving Forward 147

Index 150

ACKNOWLEDGEMENTS

This book was supported by my NSF CAREER Award (#1350068) in the U.S. (2014–2020). Many of the cases presented were adapted from actual classroom data collected for the NSF project. I thank all in-service teachers who provided the original data that made the development of the contrasting cases possible.

I am also grateful for the research support provided by the College of Education and Human Development (CEHD) at Temple University (2024–2025). This support enabled me to pilot the contrasting cases in my methods courses and to collect data from preservice teachers who provided consent for its use. The "voices" shared in this book are included with their permission, and I am thankful to all preservice teachers who participated in this project.

Moreover, with support from CEHD, VR-supported videos were developed based on four pairs of contrasting cases (Cases 1, 3, 6, and 9) through a collaboration with Dr. Di Liu and student Ms. Jingyi Guo. These videos have been made available on YouTube for broad access (links provided in the footnote of Chapter 1). I am thankful to Dr. Liu and Jingyi for their contributions.

I also appreciate Temple University's 2025 Summer Research Award from the Provost's Office. This support allowed me to dedicate my summer to completing the draft of this book. Above all, I am deeply thankful to my family, especially my husband Xiaobao, who lovingly took care of our son Owen during the summer and enabled me to focus on writing.

Throughout the development and production of this book, I am grateful to the Routledge editors—Julie Dolinger, Rosaleah Stammler, and Mia Balashova—as well as to the senior project manager at KnowledgeWorks Global Ltd, Ashraf Reza. Their delegation, patience, and timely support are inspiring and truly appreciated.

Finally, I acknowledge the use of OpenAI's ChatGPT for language editing assistance in preparing this book. Any remaining errors are my own.

1

INTRODUCTION

In this chapter, I introduce the background and rationale for this book, including how it was developed, why it is needed, and the new contributions it makes to the field. Specifically, I highlight the importance of developing preservice elementary teachers' (PSTs') knowledge and skills in mathematical planning and teaching, as well as the lack of systematic teaching support for their learning in current mathematics methods courses. This leads to the introduction of one teaching approach, TEPS, which is illustrated through ten pairs of contrasting cases and justified by the Interactive, Constructive, Active, and Passive (ICAP) framework. Finally, I describe how I piloted these contrasting cases in my math methods course (Fall 2024–Fall 2025), further supporting the development of this book.

1.1 Background

Elementary mathematics classroom teaching plays a critical role in developing students' mathematical thinking and learning. However, many classrooms focus on answer-seeking rather than understanding concepts and underlying ideas. Elementary preservice teachers in many countries, including the U.S., are often products of such a K-12 system and possess weak mathematical knowledge for teaching. This highlights the need to better equip elementary teachers with knowledge and skills in mathematical planning and instruction.

Throughout a teacher's career, in contrast to the fully packed schedule of an in-service teacher, a preservice teacher (PST) in an education program can dedicate at least a semester to learning how to teach mathematics. As such, many scholars (Ma, 1999; Morris et al., 2009) suggest seizing this opportunity to prepare teachers while they are still in their programs. Once future teachers are well prepared for in-depth teaching, their students are more likely to experience in-depth learning. The *Standards for Preparing Teachers of Mathematics* (Association of Mathematics Teacher Educators, 2017) emphasizes that well-prepared beginning teachers should have strong, connected mathematical knowledge for teaching, a deep understanding of students and their learning, and the ability to use effective, equitable instructional practices.

Although the need to prepare PSTs for in-depth mathematics teaching has been widely acknowledged, existing instructional materials for PSTs either broadly describe effective teaching (e.g., mathematics methods books) or present teaching strategies in a dispersed manner (e.g., National

DOI: 10.4324/9781003586333-1

Council of Teachers of Mathematics [NCTM] casebooks; see Schifter et al., 2016a, 2016b), and thus may not provide teachers with a coherent mental model of good teaching. This highlights the need to provide future teachers with a systematic teaching approach, illustrated with concrete lesson images or scripts, including both effective and less effective teaching cases. Through in-depth analysis of these contrasting lesson images, PSTs are likely to develop a deeper understanding of effective mathematics teaching.

1.2 Goal of the Book

The primary goal of this book is to equip future teachers with teaching expertise in elementary mathematics, though it can also serve as a useful tool for supporting in-service teachers. In particular, I present a teaching approach called *Teaching through Example-based Problem Solving* (TEPS), which is detailed in Chapter 2. The TEPS approach, described in Ding (2021), is based on insights derived from my prior National Science Foundation CAREER award (#1350068, 2014–2020), in conjunction with my accumulated research and teaching experience in elementary mathematics. In this book, the TEPS approach is illustrated through ten pairs of contrasting cases, each comparing TEPS with a conventional approach.

To help PSTs understand how and why cases using the TEPS approach provide more valuable learning opportunities than the conventional approach, I use the Interactive, Constructive, Active, and Passive (ICAP) framework (Chi, 2021; Chi et al., 2018; Chi & Wylie, 2014)—a widely used model for classifying students' cognitive engagement based on their learning behaviors—as justification. This framework is also presented in detail in Chapter 2.

1.3 Why Contrasting Cases?

The motivation for developing contrasting cases is twofold. Practically, with prior NSF support, my research team and I have collected over 100 videos of early algebra lessons taught by expert elementary teachers in the U.S. and China. These cross-cultural lessons provide valuable insights and highlight areas for improvement, as documented in my recent book (Ding, 2021). However, direct comparisons of matched cases are limited because the lessons are based on country-specific textbooks. Furthermore, due to research ethics, the video clips cannot be freely shared with the public. To address this challenge and share the identified insights with teachers, I developed contrasting cases. These draw partly from my NSF project lessons but use modified tasks (e.g., items suggested by the Common Core) and pseudonyms for teachers (e.g., Teacher X, Teacher Y).

Theoretically, contrasting cases are an effective instructional tool because they provide opportunities for comparison—a higher-order thinking process that can promote deep learning (Rittle-Johnson & Star, 2009, 2011). However, without explicit guidance, contrasting cases may not be effective, as learners might focus on surface features rather than underlying concepts (Loibl et al., 2020). Therefore, instructors need to provide meaningful guidance when implementing contrasting cases. When PSTs are guided to actively engage in comparison, they are more likely to develop a deeper understanding of effective mathematics teaching.

In this book, each case includes two paired sub-cases that take different approaches: TEPS versus a conventional approach. The main differences lie in teachers' use of representations and questioning when unpacking a worked example task, which reflect their different instructional foci. By comparing these aspects, PSTs can develop a clearer sense of what good teaching looks like. Based on the ICAP framework introduced below, PSTs are expected to reach at least the constructive level of learning through their engagement with contrasting cases.

1.4 Why Early Algebra?

The mathematical content highlighted in the contrasting cases focuses on two fundamental mathematical ideas—also called "early algebra": Inverse relations and the basic properties of operations. Inverse relations refer to the relationships between addition and subtraction (known as the "additive inverse") and between multiplication and division (known as the "multiplicative inverse"). In the U.S., additive inverses are typically introduced in first and second grade, while multiplicative inverses are introduced in third and fourth grade. Researchers (Nunes et al., 2009) argue that students cannot fully comprehend the four arithmetic operations without an understanding of inverse relations. Furthermore, a solid grasp of inverse relations is a strong predictor of students' algebraic learning (Stern, 2005).

Similarly, the basic properties of operations are considered "backbones" to algebraic thinking (Howe, 2014; Wu, 2009). In elementary school, the basic properties refer to the commutative, associative, and distributive properties, often known as the laws of arithmetic. Flexible use of these properties can simplify computation, demonstrating both the power of mathematical properties and the inherent beauty of mathematics.

Since algebra is widely recognized as a gatekeeper for students' overall mathematics learning, developing algebraic thinking in elementary school has attracted significant attention. However, prior research (Barnett & Ding, 2019; Ding et al., 2019, 2023) indicates that many elementary teachers often struggle to foster students' algebraic thinking. To address this, the current book presents ten pairs of contrasting cases focused on early algebra: Inverse relations (Cases 1–5) and the basic properties of operations (Cases 6–10). Each case contains two contrasting lesson images (detailed lesson scripts) or sub-cases—one illustrating the TEPS approach and the other demonstrating a conventional approach commonly seen in elementary classrooms. Through these contrasting cases, PSTs are expected to understand the TEPS approach and develop the expertise necessary to teach elementary mathematics effectively.

1.5 Pilot Work

During the Fall 2024 and Fall 2025 semesters, with our college's support, I piloted these contrasting cases with PSTs in my mathematics methods course at Temple University. At the beginning of the semester, PSTs were asked to read and respond to Chi and Boucher's (2023) book chapter on the ICAP framework. Over the following months, three Contrasting Cases Analyses (CCA) were conducted—one per month. PSTs read and annotated the selected contrasting cases, discussed their similarities and differences in small groups, and wrote reports individually or collaboratively. The discussion questions are shared in the later chapters. Additionally, PSTs completed a Lesson Plan Project based on their chosen topics, and their lesson plans were analyzed using ICAP. Beyond these assignments, I conducted ongoing surveys with PSTs to gather their feedback on the effectiveness of the assignments and maintained logs and reflections on using the CCA.

Although the primary aim of this book is not to document the effect of contrasting cases on PSTs' learning, I include some PSTs' voices, as well as my ongoing observations and reflections. These insights may be valuable for math educators who wish to use this book as a supplementary resource to support PSTs' learning.

Additionally, as an extension of this book, my research team and I have developed four pairs of VR-supported videos in which contrasting cases are read aloud and acted out.[1] These VR-supported videos may provide additional support for methods course instructors, who can play the videos along with the written scripts. My pilot work reveals that, although the video quality could be improved, most PSTs appreciated watching the videos alongside the written cases.

1.6 Structure of the Book

In the following chapters, I first present the two analytical frameworks, TEPS and ICAP (Chapter 2). Next, I present the contrasting cases in order of additive inverse (Chapter 3), multiplicative inverse (Chapter 4), basic properties of addition (Chapter 5), and basic properties of multiplication (Chapter 6). I revisit the TEPS model, contrasting cases, and ICAP in the conclusion (Chapter 7).

Note

1 The development of the VR videos was supported by the College of Education and Human Development's (CEHD) strategic fund at Temple University, with collaborative contributions from Dr. Di Liu and Ms. Jingyi Guo. The YouTube playlists of videos are as follows:

- Playlist of Zoom Recordings: https://youtube.com/playlist?list=PL6gRZY_sWmmo1BH_BJQLjsxBN4za-N1dt&si=bU-q2PlfcziCi35x
- Playlist of VR Recordings: https://youtube.com/playlist?list=PL6gRZY_sWmmoIs4OtsZZuQIUAvv3R2ohD&si=QbQByYHykdNqcyTN

References

Association of Mathematics Teacher Educators. (2017). *Standards for preparing teachers of mathematics*. https://amte.net/standards.

Barnett, E., & Ding, M. (2019). Teaching of the associative property: A natural classroom investigation. *Investigations in Mathematics Learning, 11*, 148–166. https://doi.org/10.1080/19477503.2018.1425592

Chi, M. T., & Boucher, N. S. (2023). Applying the ICAP framework to improve classroom learning. In C. E. Overson et al. (Eds.). *In their own words: What scholars want you to know about why and how to apply the science of learning in your academic setting* (pp. 94–110). American Psychological Association.

Chi, M. T. H. (2021). Translating a theory of active learning: An attempt to close the research-practice gap in education. *Topics in Cognitive Science, 13*(3), 441–463. https://doi.org/10.1111/tops.12539

Chi, M.T.H., Adams, J., Bogusch, E.B., Bruchok, C., Kang, S., Lancaster, M., Levy, R., et al. (2018). Translating the ICAP theory of cognitive engagement into practice. *Cognitive Science, 42*, 1777–1832. https://doi.org/10.1111/cogs.12626

Chi, M. T. H., & Wylie, R. (2014). The ICAP framework: Linking cognitive engagement to active learning outcomes. *Educational Psychologist, 49*, 219–243. https://doi.org/10.1080/00461520.2014.965823

Ding, M. (2021). *Teaching early algebra through example-based problem solving: Insights from Chinese and U.S. elementary classrooms*. Routledge. https://www.routledge.com/Teaching-Early-Algebra-through-Example-Based-Problem-Solving-Insights-from/Ding/p/book/9780367431785

Ding, M., Chen, W., & Hassler, R. (2019). Linear quantity models in US and Chinese elementary mathematics classrooms. *Mathematical Thinking and Learning, 21*, 105–130. https://doi.org/10.1080/10986065.2019.1570834

Ding, M., Li, X., Manfredonia, M., & Luo, W. (2023). US and Chinese elementary teachers' noticing of cross-cultural mathematics videos. *Journal of Mathematics Teacher Education, 26*, 211–239. https://doi.org/10.1007/s10857-021-09526-z

Howe, R. (2014). Three pillars of first grade mathematics, and beyond. In Y. Li & G. Lappan (Eds.). *Mathematics curriculum in school education: Advances in mathematics education*. Springer.

Loibl, K., Tillema, M., Rummel, N., & van Gog, T. (2020). The effect of contrasting cases during problem solving prior to and after instruction. *Instructional Science*, *48*, 115–136. https://doi.org/10.1007/s11251-020-09504-7

Ma, L. (1999). *Knowing and teaching elementary mathematics: Teachers' understanding of fundamental mathematics in China and the United States*. Routledge.

Morris, A. K., Hiebert, J., & Spitzer, S. M. (2009). Mathematical knowledge for teaching in planning and evaluating instruction: What can preservice teachers learn? *Journal for Research in Mathematics Education, 40*(5), 491–529. https://www.jstor.org/stable/40539354

Nunes, T., Bryant, P., & Watson, A. (2009). *Key understandings in mathematics learning: A report to the Nuffield Foundation*. Nuffield Foundation. https://www.nuffieldfoundation.org/project/key-understandings-in-mathematics-learning

Rittle-Johnson, B., & Star, J. R. (2009). Compared with what? The effects of different comparisons on conceptual knowledge and procedural flexibility for equation solving. *Journal of Educational Psychology*, *101*(3), 529. https://doi.org/10.1037/a0014224

Rittle-Johnson, B., & Star, J. R. (2011). The power of comparison in learning and instruction: Learning outcomes supported by different types of comparisons. In J. P. Mestre & B. H. Ross (Eds.), *The psychology of learning and motivation: Cognition in education* (pp. 199–225). Elsevier Academic Press. https://doi.org/10.1016/B978-0-12-387691-1.00007-7

Schifter, D., Bastable, V., & Russell, S. J. (2016a). *Number and operations, part 1: Building a system of tens.* Casebook. *National Council of Teachers of Mathematics*.

Schifter, D., Bastable, V., & Russell, S. J. (2016b). *Number and operations, part 2: Making meaning for operations casebook. National Council of Teachers of Mathematics*.

Stern, E. (2005). Knowledge restructuring as a powerful mechanism of cognitive development: How to lay an early foundation for conceptual understanding in formal domains. *British Journal of Educational Psychology*, *2*(3), 155–170. http://doi.org/10.1348/000709905X62453

Wu, H. (2009). *Learning algebra*. http://math.berkeley.edu/~wu/Bethel-5.pdf

2

ANALYTICAL FRAMEWORKS

TEPS and ICAP

In this chapter, I introduce two analytical frameworks that guide the development and analysis of the contrasting cases in this book. The first framework, *Teaching through Example-based Problem Solving* (TEPS), is documented in my prior research book (Ding, 2021) and illustrated by contrasting cases developed in the current book. To analyze and justify the contrasting cases, the second framework, ICAP, is introduced. This framework identifies four levels of students' cognitive engagement—Interactive, Constructive, Active, and Passive—leading to either deep or shallow learning (Chi et al., 2018). Both frameworks are applied throughout this book to help readers understand the distinctions between the contrasting cases and their significance.

2.1 TEPS: Teaching through Example-based Problem Solving

The development of TEPS is guided by the Institute of Education Sciences (IES) recommendations (Pashler et al., 2007) and enriched by my prior international comparative research on Chinese and U.S. expert teachers' classroom practices (Ding, 2021). Overall, TEPS emphasizes in-depth unpacking of a worked example through problem-solving. Hence, the cases introduced in this book focus mainly on worked example sections. The TEPS approach takes a middle ground in the ongoing debate between worked examples and problem-solving approaches (Anderson et al., 2000). While TEPS emphasizes the worked example effect (Sweller & Cooper, 1985), it does not simply introduce a worked example for students to study passively. At the same time, while it encourages student autonomy, it does not ask students to solve problems themselves without guidance. Rather, teachers engage students in solving an example problem to develop mental schemas, enabling independent problem solving.

To promote deep unpacking of a worked example, TEPS emphasizes two pedagogical aspects: the use of representations and deep questioning. These aspects are often intertwined, as teachers typically ask questions about representations. Elaboration follows.

Representation Uses. In a math classroom, teachers may use varied representations, ranging from concrete (e.g., real-world objects or contexts) to semi-concrete (e.g., manipulatives or sketches) to abstract (e.g., numbers or definitions). These representations can be sequenced in different ways, either from concrete to abstract or from abstract to concrete. Research strongly

DOI: 10.4324/9781003586333-2

supports the first sequence, known as *concreteness fading*, which gradually transitions from concrete to abstract representations, promoting learning and transfer (Fyfe et al., 2015; Ottmar & Landy, 2017). This approach is advocated by some educational psychologists (e.g., Fyfe & Nathan, 2019) and has been observed empirically in mathematically high-achieving East Asian countries (Ding & Li, 2014; Ding et al., 2019; Murata, 2008; Ng & Lee, 2009). In TEPS, I recommend situating a worked example in a real-world context relevant to children's experiences to enhance engagement. Teachers should then guide students in modeling the real-world situation, following the concreteness fading sequence.

Deep Questions. Teachers should also ask deep questions to elicit students' explanations of targeted concepts. Although non-deep questions are needed in a lesson, the absence of deep questions signals a lack of deep learning. While students are often expected to provide explanations, they may not do so spontaneously. Therefore, teachers need to ask carefully crafted, deep questions to promote students' meaningful explanations. Deep questions often begin with stems such as *why, how, what if, what if not*, or *how does X compare to Y* (Craig et al., 2006; Pashler et al., 2007). These questions prompt students to go beyond existing knowledge and generate new inferences. In TEPS, two types of deep questions are highlighted: Concept-specific questions, which promote meaning-making of particular aspects, and comparison questions, which encourage connections across multiple aspects to support deeper understanding (Ding, 2021). Of course, there are other types of deep questions. For instance, a teacher may invite students to pose questions or problems (e.g., Can you pose a question?). Posing questions or problems involves complex processes of problem generation and reflection (Cai & Rott, 2024), and it is widely considered a valuable way to foster students' creative thinking and problem-solving skills (Cai, 2022). In our crafted cases, readers will also encounter problem-posing questions.

The TEPS approach differs from conventional teaching practices. In terms of representations, many teachers start with abstract definitions or numerical examples rather than following a concreteness fading sequence. Real-world contexts, if used, often serve only as a pretext for computation, overlooking the underlying quantitative relationships (Ding & Carlson, 2013; Ding et al., 2023). Regarding questioning, conventional approaches tend to rely on show-and-tell or computationally focused questions, which do not support students' mathematical thinking and learning. In this book, I analyze the contrasting cases with a focus on teachers' use of representations and questioning strategies, highlighting the key features of TEPS and conventional practices for comparison, and justifying these analyses through the ICAP framework (see 2.2).

2.2 ICAP: Interactive, Constructive, Active, and Passive

To justify why the TEPS approach is more effective, I refer to the highly regarded ICAP cognitive engagement framework. ICAP is a learning theory developed by Chi and her team based on analyses of hundreds of studies (Chi et al., 2018). The ICAP framework has been successfully applied across disciplines, including psychology, science, engineering, and medical education, as a tool for measuring lesson quality and student engagement (Chi et al., 2017; Menekse & Chi, 2019; Morris & Chi, 2020).

According to Chi and colleagues (Chi, 2009, 2021; Chi & Wylie, 2014; Chi et al., 2018), students' learning occurs at four levels—Interactive, Constructive, Active, and Passive—ranked in order of expected learning outcomes (I > C > A > P). Students who passively receive information (P level) primarily store it in isolation, without activating prior knowledge or making meaningful connections (Chi et al., 2018). For example, a teacher defining a mathematical term and asking students to read it aloud reflects passive engagement. At the Active (A) level, students manipulate information but do not draw new inferences beyond the given material. For instance, using manipulatives such as cubes without linking them to broader mathematical concepts represents limited cognitive engagement. Consequently, both P and A levels involve minimal deep thinking, often resulting in shallow learning.

In contrast, Constructive (C) and Interactive (I) activities require students to generate or co-generate new information, fostering deeper cognitive engagement and learning (Chi, 2009; Chi & Wylie, 2014). For example, when teachers prompt students to compare mathematical solutions or analyze differences, they encourage reasoning beyond the given materials. Interactive (I) engagement goes beyond Constructive (C) by requiring collaboration, such as small-group discussions in which students co-generate new ideas. However, group work alone does not guarantee Interactive engagement; without effective teacher guidance, discussions may remain superficial, defaulting to the A level (Chi & Boucher, 2023).

The ICAP framework builds on established learning theories, including constructivism, Bloom's taxonomy, and cognitive load theory, but distinguishes itself by focusing on students' overt or observable learning behaviors and, where applicable, their tangible outputs (Chi et al., 2018). As Chi and Boucher (2023) note, ICAP provides simple yet powerful heuristics for differentiating among various forms of "active learning" and assessing their likely impact.

In this book, I highlight how teaching elements in different approaches (TEPS vs. conventional) can influence students' cognitive engagement levels (I > C > A > P), leading to varying depths of learning, from shallow to deep. Note that this book is intended for practitioners. The ICAP model is used mainly as a heuristic, applied to lessons as a whole rather than to a detailed analysis of individual teaching moves. Additionally, a direct comparison of the contrasting cases is also expected to help preservice teachers (PSTs) recognize how missed opportunities in conventional lessons could be enhanced by incorporating insights from the TEPS approach. Ultimately, my goal is for teachers to apply this approach in their own classrooms, effectively teaching early algebra and beyond while fostering high levels of student cognitive engagement. I believe that the targeted TEPS approach can empower teachers to optimize instructional tasks, regardless of the elementary mathematics textbooks they use.

References

Anderson, J. R., Reder, L. M., & Simon, H. A. (2000, Summer). Applications and misapplications of cognitive psychology to mathematics education. *Texas Educational Review*. https://act-r.psy.cmu.edu/papers/misapplied.html

Cai, J. (2022). What research says about teaching mathematics through problem posing. *Éducation et Didactique*, *16*(3), 31–50. https://doi.org/10.4000/educationdidactique.10642

Cai, J., & Rott, B. (2024). On understanding mathematical problem-posing processes. *ZDM – Mathematics Education*, *56*(1), 61–71. https://doi.org/10.1007/s11858-023-01536-w

Chi, M. T., & Boucher, N. S. (2023). Applying the ICAP framework to improve classroom learning. In C. E. Overson et al. (Eds.). *In their own words: What scholars want you to know about why and how to apply the science of learning in your academic setting* (pp. 94–110). American Psychological Association.

Chi, M. T. H. (2009). Active-constructive-interactive: A conceptual framework for differentiating learning activities. *Topics in Cognitive Science*, *1*, 73–105. https://doi.org/10.1111/j.1756-8765.2008.01005.x

Chi, M. T. H. (2021). Translating a theory of active learning: An attempt to close the research-practice gap in education. *Topics in Cognitive Science*, *13*, 441–463. https://doi.org/10.1111/tops.12539

Chi, M. T. H., Adams, J., Bogusch, E. B., Bruchok, C., Kang, S., Lancaster, M., & Levy, R. et al. (2018). Translating the ICAP theory of cognitive engagement into practice. *Cognitive Science*, *42*, 1777–1832. https://doi.org/10.1111/cogs.12626

Chi, M. T. H., Kang, S., & Yaghmourian, D. L. (2017). Why students learn more from dialogue- than monologue-videos: Analyses of peer interactions. *Journal of the Learning Sciences*, *26*, 10–50. https://doi.org/10.1080/10508406.2016.1204546

Chi, M. T. H., & Wylie, R. (2014). The ICAP framework: Linking cognitive engagement to active learning outcomes. *Educational Psychologist*, *49*, 219–243. https://doi.org/10.1080/00461520.2014.965823

Craig, S. D., Sullins, J., Witherspoon, A., & Gholson, B. (2006). The deep-level-reasoning-question effect: The role of dialogue and deep-level-reasoning questions during vicarious learning. *Cognition and Instruction*, *24*, 565–591.

Ding, M. (2021). *Teaching early algebra through example-based problem solving: Insights from Chinese and U.S. Elementary classrooms*. Routledge.

Ding, M., & Carlson, M. A. (2013). Elementary teachers' learning to construct high quality mathematics lesson plans: A use of IES recommendations. *The Elementary School Journal*, *113*, 359–385. https://doi.org/10.1086/668505

Ding, M., Chen, W., & Hassler, R. (2019). Linear quantity models in US and Chinese elementary mathematics classrooms. *Mathematical Thinking and Learning*, *21*, 105–130. https://doi.org/10.1080/10986065.2019.1570834

Ding, M., & Li, X. (2014). Transition from concrete to abstract representations: The distributive property in a Chinese textbook series. *Educational Studies in Mathematics*, *87*, 103–121. https://doi.org/10.1007/s10649-014-9558-y

Ding, M., Li, X., Manfredonia, M., & Luo, W. (2023). US and Chinese elementary teachers' noticing of cross-cultural mathematics videos. *Journal of Mathematics Teacher Education, 26*, 211–239. https://doi.org/10.1007/s10857-021-09526-z

Fyfe, E. R., McNeil, N. M., & Borias, S. (2015). Benefits of "concreteness fading" for children's mathematics understanding. *Learning and Instruction*, *35*, 104–120.

Fyfe, E. R., & Nathan, M. J. (2019). Making "concreteness fading" more concrete as a theory of instruction for promoting transfer. *Educational Review*, *71*(4), 403–422. https://doi.org/10.1080/00131911.2018.1424116

Menekse, M., & Chi, M. T. H. (2019). The role of collaborative interactions versus individual construction on students' learning of engineering concepts. *European Journal of Engineering Education*, *44*, 702–725. https://doi.org/10.1080/03043797.2018.1538324

Morris, J., & Chi, M. T. H. (2020). Improving teacher questioning in science using ICAP theory. *The Journal of Educational Research*, *113*(1), 1–12. https://doi.org/10.1080/00220671.2019.1709401

Murata, A. (2008). Mathematics teaching and learning as a mediating process: The case of tape diagrams. *Mathematical Thinking and Learning*, *10*(4), 374–406.

Ng, S. F., & Lee, K. (2009). The model method: Singapore children's tool for representing and solving algebraic word problems. *Journal for Research in Mathematics Education*, *40*, 282–313.

Ottmar, E., & Landy, D. (2017). Concreteness fading of algebraic instruction: Effects on learning. *Journal of the Learning Sciences*, *26*(1), 51–78. https://doi.org/10.1080/10508406.2016.1250212

Pashler, H., Bain, P. M., Bottge, B. A., Graesser, A., Koedinger, K., & McGaniel, M. et al. (2007). *Organizing instruction and study to improve student learning* (NCER 2007-2004). National Center for Education Research.

Sweller, J., & Cooper, G. A. (1985). The use of worked examples as a substitute for problem solving in learning algebra. *Cognition and Instruction*, *2*(1), 59–89. https://doi.org/10.1207/s1532690xci0201_3

3

TEACHING ADDITIVE INVERSES

Contrasting Cases 1–2

In this chapter, I present two cases (Cases 1 and 2) on teaching additive inverse using different models: part-whole and comparison, respectively. Each case contains two contrasting sub-cases that illustrate different instructional approaches in terms of teachers' use of representations and deep questioning.

In the TEPS approach (Cases 1a and 2a), the example tasks are situated in real-world contexts (inverse word problems suggested by the Common Core). Tape diagrams are used to illustrate quantitative relationships and to support the transition from concrete to abstract solutions. Deep questions are raised to promote students' problem-posing, meaning-making, and comparisons, eliciting their structural understanding of inverse relations. According to ICAP, these inference-making opportunities are likely to engage students in deep learning—at the constructive and interactive levels—of the targeted concept.

In contrast, in the conventional approach (Cases 1b and 2b), teaching focuses primarily on computational answers. The representational sequence is opposite to concreteness fading, often starting with abstract number sentences or semi-concrete representations (e.g., cubes). Word problems may or may not be presented; if included, they typically appear at the end of the lesson as a pretext to apply the learned computation. Due to its computational focus, deep questions are rarely posed. When potential deep questions arise, anticipated responses remain at a surface rather than structural level, requiring little to no inference-making. In this sense, students' cognitive engagement is likely to remain at the passive or, at most, active level.

To engage readers, I adopt a three-step process for presenting each case hereafter. First, I invite readers to "think and prepare" (Step 1) by generating an initial draft plan for the same topic. Second, readers are invited to "read and annotate" (Step 2) the presented cases. Third, after reading and annotating the cases, readers are asked to "stop and reflect" (Step 3) based on four questions focusing on representation use, deep questioning, ICAP level, and insights gained. In my math methods course, I designated "Step 1" as PSTs' preparation work before class and conducted "Step 2" during class. "Step 3" was initiated in class through individual reflection and group discussions and completed afterward as a written report, either in narrative or tabular form. Below, I present each pair of contrasting cases, followed by the key to analysis based on the first three questions. For the last question, insights may vary; I share common responses from my preservice teachers based on pilot work.

DOI: 10.4324/9781003586333-3

3.1 Teaching Additive Inverses with Part-whole Model

3.1.1 Introduction

The Common Core State Standards expect students in grades K-2 to understand the relationship between addition and subtraction. For example, students should understand how $5 + 2 = 7$ is related to $7 - 2 = 5$. This understanding can be introduced in kindergarten through various representations (writing equations is encouraged but not required), formally discussed in Grade 1, and extended to larger numbers (within 100) in Grade 2. In Case 1, I will discuss how inverse relations can be illustrated through part-whole word problems.

STEP 1: THINK AND PREPARE

Imagine you are a first-grade teacher. How would you teach inverse relations (e.g., $3 + 2 = 5$, $2 + 3 = 5$, $5 - 3 = 2$, $5 - 2 = 3$) based on a part-whole model? Draft a brief plan in a T-S dialogue format. What representations would you use, and how would you sequence them? What questions, including deep questions, would you ask, and what student explanations would you anticipate?

Your Initial Draft Plan:

3.1.2 Contrasting Cases: 1a and 1b

In Case 1 below, there are two subcases illustrating how two first-grade teachers approach a worked example of inverse relations (not a full lesson). Note that throughout this book, the lesson scripts are presented in a T–S dialogue format with emphasis on teachers' moves[1]. Case 1a features Teacher X using the TEPS approach, while Case 1b showcases Teacher Y using a conventional method. Both lessons involved a set of inverse word problems of part-whole model, adapted from the Common Core State Standards for Mathematics (see Table 3.1).

Note that students in both classes possess the same prior knowledge, and the two lessons share the same teaching objectives:

Prior Knowledge: Students have already learned how to solve basic addition and subtraction problems, but they have not yet explored the relationships between these operations.

Teaching Objectives: Students will learn how addition and subtraction are related to each other based on the part–whole model.

TABLE 3.1 Part-whole Inverse Word Problems Adapted from the Common Core

Whole Unknown	*Part Unknown*	*Part Unknown*
Three red apples and two green apples are on the table. How many apples are on the table? $(3 + 2 = 5)$	Five apples are on the table. Three are red and the rest are green. How many apples are green? $(5 - 3 = 2)$	Five apples are on the table. Two are green and the rest are red. How many apples are red? $(5 - 2 = 3)$

STEP 2: READ AND ANNOTATE

Below are cases 1a and 1b. Read and annotate each case, paying particular attention to the use of representations and questioning. Specifically, for each teacher (X and Y), identify:

a The types of representations used
b The deep questions asked

Case 1a: Teacher X

1. Present a picture about apples

T: Do you like apples, friends? (Yes!) (Shows the picture of apples, including a mix of red and green apples. See Figure 3.1.) What do you see in this picture? (3 red apples and 2 green apples.)

T: Yes, we see "Three red apples and two green apples are on the table." Can you ask a question about this picture?

[Possible student responses: (1) How many more red apples are there than green apples? (2) How many apples are on the table?]

T: (Praises the first response and tells students that they will study it in the future). Let's focus on the second question. (Read the word problem completely):

There are 3 red apples and 2 green apples on the table. How many apples are on the table?

2. Discuss the two addition word problems

T: If we use red cubes to represent red apples and green cubes to represent green apples, can you show this story problem on your desk? Two of you at each desk can work together.

(Ask one pair of students to demonstrate on the board. Project their cubes so that the class can follow along and explain the solution.)

T: Here is what I did, similar to your cubes (see Figure 3.2).

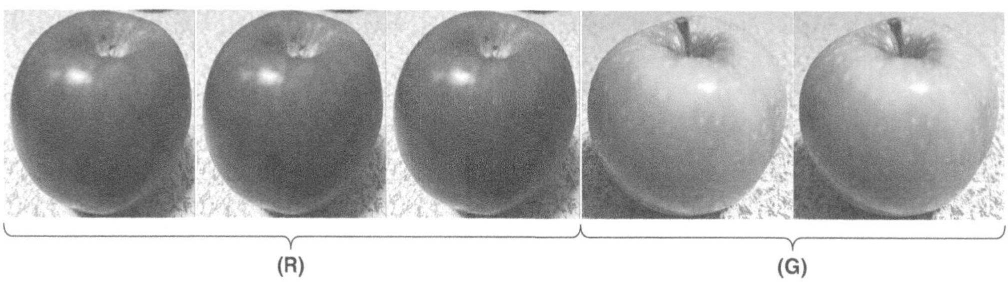

FIGURE 3.1 Picture of apples

Note: Due to black-and-white printing, the labels **R** and **G** indicate the colors of the objects. This notation is used throughout.

FIGURE 3.2 The cube representation for 3 red and 2 green apples

T: To find out how many apples there are in total, what equation can we write? (3 + 2 = 5)

T: Can you explain the meaning of each number in this equation? (3 red apples, 2 green apples, 5 total apples).

T: (Box the red and green counters, respectively). We call the red apples one "part", the green apples the other "part," and the total apples the "whole."

(Write "part," "part," and "whole" on the board and ask students to read them aloud. See Figure 3.3.)

T: In the above problem, we used 3 + 2 to solve it. Why did we add? (Because we know the two parts and need to find the whole).

T: Great. Do we have another way to solve this story problem?

- If needed, remind students that in 3 + 2 = 5, we first looked at the "red" apples. Now what color can we look at first? What equation would that give to us? (2 + 3 = 5).
- Can you explain the meaning of 2 + 3 = 5 using your cubes? (Put 2 green apples and 3 red apples together; I will get 5 apples in total).

T: Compare these two addition equations. What do you find? How are they similar and different? Share with your deskmate first.

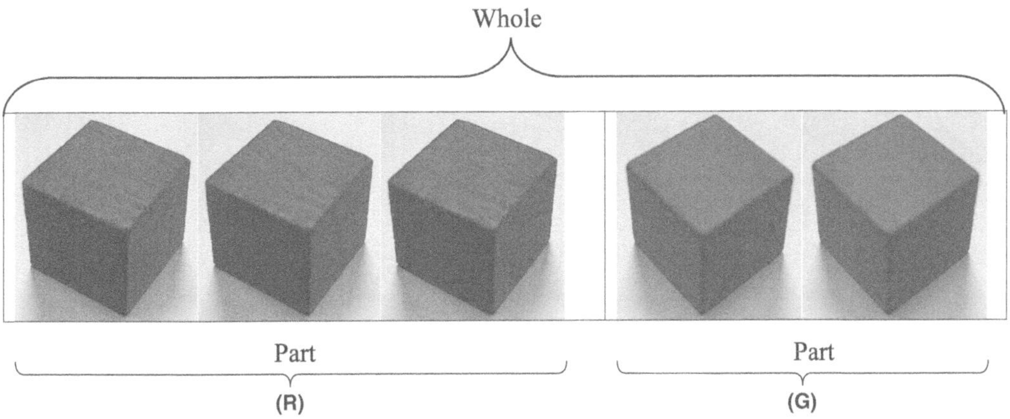

FIGURE 3.3 The cube representation, labeled with "part," "part," and "whole"

(Students should eventually see that 3 and 2 are switched, but the answer remains the same).

T: Why do 3 + 2 and 2 + 3 both equal 5? Can we explain this using either your cubes or these pictures?

(Students should eventually see that although we switched the two parts around, the total does not change. The teacher should explicitly summarize this main point).

3. Discuss the two subtraction word problems

T: (Pointing to the cube picture) We just used two ways to find the total number of apples. What if we already know there are 5 apples, which include both red and green? (Write "5" and cover the green cubes). What do we see now? (see Figure 3.4).

S: There are three red apples!

T: Can we pose a question based on what we already know?

- Ask students to say it completely: *Five apples are on the table. Three are red and the rest are green. How many apples are green?*

T: How to solve it? (5 – 3 = 2). Why did we use subtraction?

- Allow multiple students to explain. The teacher then emphasizes: "*When we know the whole and one part, and need to find the other part, we use subtraction. In other words, when we take away one part from the whole, we find the other part.*"

T: What if we still have 5 apples and I cover the red part of apple (see Figure 3.5). Who can come up with a word problem? (Encourage students to verbalize it.)

T: How to solve it? (5 – 2 = 3). Can you explain the meaning of your equation? (We have 5 apples, 2 are green. Three are red.)

T: Compare these two subtraction equations. What do you notice? (Both start with the whole "5," but each takes away a different part).

FIGURE 3.4 Cube representation for the first subtraction task

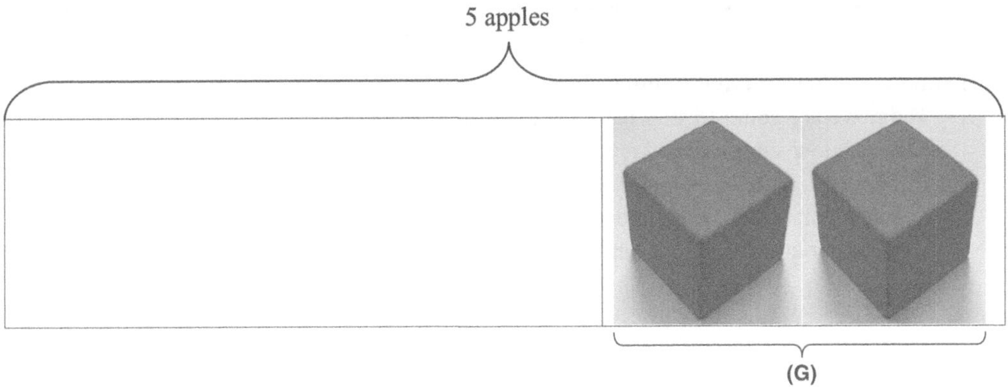

FIGURE 3.5 Cube representation for the second subtraction task

4. Discuss the four equations

T: We just obtained a group of equations.

$3+2=5$

$2+3=5$

$5-2=3$

$5-3=2$

T: Two of these are addition and two are subtraction. In small groups, can you explain the meaning of each equation based on the story problem? You can use your cubes to help.

T: Let's compare these four equations only. What do you notice?

(Encourage students to verbalize their observations of similarities and differences.)

T: Have the meanings of 3, 2, and 5 in each equation changed? (No. They're in a different order, but it's still the same part, part, and whole). We call such a group of related equations a "fact family."

- Why do we call it a fact family? (Because in this group, the addition and subtraction facts are related to each other).
- When do we use addition, and when do we use subtraction? (To find the whole, we use addition; to find a part, we use subtraction).

5. Summarize the inverse relationship using a part-whole diagram/tape diagram

T: Great observations! (Use the counter picture on the board to help explain). Today, we have learned about a group of facts or problems that share the same parts and the same whole. To find the whole, we use…? (Addition; part + part = whole).

To find the part, we use…? (Subtraction; whole – part = part).

Let's take our hands to show how addition and subtraction are related. (Say and gesture together: Part + part = whole; whole – part = part).

Case 1b: Teacher Y

1. Start with three related number facts

T: Today we are going to start learning about subtraction. Knowing our addition facts will help us know subtraction.

T: I'm going to write one of your addition fact 3 + 2 = ? up here. If you know that, let's take a look at what else you should know.

(Teacher adds "5 – 3 = " and "5 – 2 = " underneath).

3+2=

5−2=

5−3=

T: What is the answer to 3 + 2 = ? (5)

T: How does knowing 3 + 2 = 5 help us solve these subtraction facts? Talk to your partners.

- S: 3 + 2 is 5, and I know if 5 is 2 bigger than 3, then 5 – 2 is 3.
- S: It all used the same numbers; it only used three numbers.

T: Okay what are those three numbers? (3, 2, and 5).

T: So, it uses these three numbers: 3, 2, and 5. Does anyone know what we call that? What is that called? (A fact family)

T: Yes, it's sometimes called a fact family. So, knowing that 3 + 2 = 5 helps you with knowing the other facts in that fact family.

2. Use cubes to help with the 4th number fact (addition)

T: I have these red and green cubes. How does this represent what I did on the board (see Figure 3.6)?

S: There are 3 cubes, and then 2 more, and it is 5.

T: Good. I don't have another addition fact family up here. Can somebody show me how you can write another addition fact using these same exact cubes? I'm not going to change them or take them apart. I'm not going to change the colors. So, what can we do to stick with the same family without changing things?

S: 2+3

FIGURE 3.6 Three red and two green cubes

FIGURE 3.7 The cube representation has been switched

T: (Pointing to the cubes) So, what can I do with this to show? This represents 3 + 2.
S: Switch them around.
T: Exactly. We can switch them around (see Figure 3.7).
T: We didn't take any cubes away or break any apart; we just switched their positions. So, what's the addition fact that would go with this? (2 + 3 = 5).
T: Have we taken any cubes away? (No.) Have we changed the numbers that are there? (No.) Very good.

3. Use cubes to represent and solve two subtraction facts

T: (pointing to the 2nd number sentence) Now, let's look at the subtraction 5 – 2. Who can use the cubes to represent 5 – 2 and tell me what the result is?. (Ask one student to come to the board).
T: (Student took away 2 green cubes). So, we started with 5 cubes and took away 2. What are we left with? (S: 3).
T: Correct. Did we add any more cubes to it? (No).
T: I'm going to put those back on. And who can show me 5 – 3 using the cubes? Remember, you will use the same cube train. (Ask another student to come to be on the board).
S: I have 5 cubes and I take away 3 (Student took away 2 green cubes and 1 red cube. See Figure 3.8).

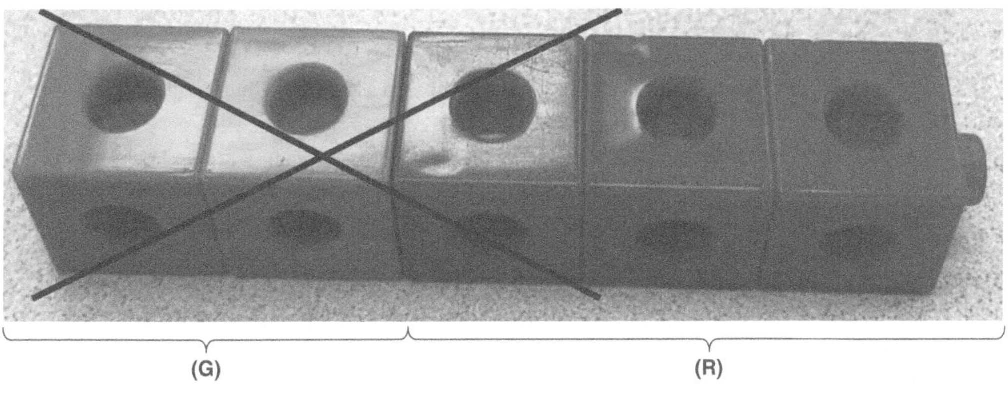

FIGURE 3.8 Student's way to take away 3 cubes

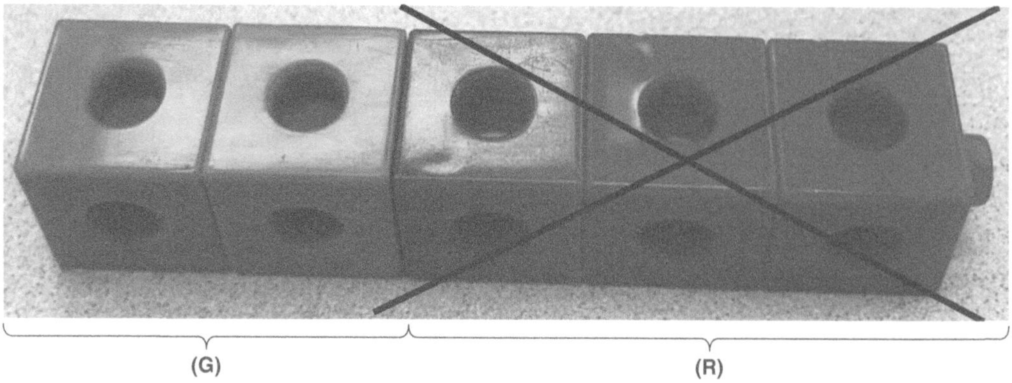

FIGURE 3.9 The student took away three red cubes, as the teacher expected

T: Okay where's the 3? We need to keep the same colors (Student was confused). If the 3 were the same color, how can you show me? (Students were still confused). Can you take away all the same color for the 3? (Student did it. See Figure 3.9.) There you go!

T: Very good, so we have not done any magic by adding anything; everything stayed the same. Thank you. So now we have our fact family of 3, 2, and 5, and we've created four facts.

4. Apply fact family to solve word problems

T: Now that we've learned about fact families, we can use that knowledge to help us solve word problems. I am going to present three problems on the board:

1 *There are 3 red apples and 2 green apples on the table. How many apples are on the table?*
2 *There are 5 apples on the table. Among them, 3 are red and the rest are green. How many apples are green?*
3 *There are 5 apples on the table. Among them, 2 are green and the rest are red. How many apples are red?*

T: In your small groups, please solve each of the word problems.

T: Let's check the solution for each word problem. (The class shared their equations: $3 + 2 = 5$, $5 - 3 = 2$, and $5 - 2 = 3$).

T: Let's compare these three equations, what do you notice? (S: They are a fact family!)

T: Exactly. These problems form a fact family, just like we discussed and modeled with our cubes! We've found the same fact family in the word problems! Great job!

5. A summary about fact family

T: Today we've learned about a fact family! Can someone tell me what a fact family is and what features it has?

S: They have the same three numbers. They are the facts from in the same family! We can use one fact to figure out the others.

Whole 5	
Part 2	Part 3

FIGURE 3.10 The part-part-whole model

T: Correct! We can use a part-part-whole model to organize a fact family (see Figure 3.10). Let's put the biggest number on the top; that is the whole. The two smaller numbers go underneath, and together they make the whole. Read after me: part-part-whole.

STEP 3: STOP AND REFLECT

Based on your reading and annotating Cases 1a and 1b, please reflect on the following:

1 **Representation uses.** Compare the representation uses of Teachers X and Y. How do their representational sequences differ? How are real-world contexts used differently?
2 **Deep Questioning.** Compare the deep questions asked by Teachers X and Y. How do they differ in terms of frequency and purpose of using deep questions?
3 **Linking to ICAP.** Based on the ICAP framework, which overall level of student engagement does each teacher likely promote? How might each teacher's lesson lead to the respective level?
4 **Insights Gained.** Compare your initial plan with the contrasting cases. What new insights have you gained that you can apply to your lesson planning and future teaching?

3.1.3 Key to Analysis of Case 1

3.1.3.1 An Overview

Cases 1a and 1b share similarities. Both lessons aim to teach the concept of additive inverse relations through a part-whole model (indicated by a fact family, $3 + 2 = 5$, $2 + 3 = 5$, $5 - 2 = 3$, $5 - 3 = 2$, involving the common core story problems). However, it appears that the lesson foci are different as indicated by teachers' representation uses and questioning. Table 3.2 below listed key points and example deep questions, followed by elaborations.

3.1.3.2 Representation Uses

Case 1a began with a real-world context—red and green apples—represented first with semi-concrete cubes and then with abstract number sentences, demonstrating a *concreteness fading* sequence. In contrast, Case 1b began with abstract number sentences, which were then illustrated using cubes. Real-world contexts were introduced only at the end, primarily to apply the learned computation skills. This sequence reverses concreteness fading.

TABLE 3.2 Key Points and Example Deep Questions for Cases 1a and 1b

	Case 1a	*Case 1b*
Representation uses	Real-world context → Semi-concrete cubes → Abstract number sentences. This sequence shows *concreteness fading*. The real-world context was used to situate the worked example and to support sense-making.	Abstract number sentences → Semi-concrete cubes → Real-world context. This sequence is opposite to concreteness fading. The real-world context was used at the end as an application of the learned computation skills.
Deep questioning	**Problem-Posing Questions** • Can you ask a question? • Can we pose a question based on what we already know? • What if we still have 5 apples and I cover the red part of apple. Who can come up with a word problem?	**Problem-Posing Questions** • NA
	Meaning-making Questions • Why did we add? Can you explain the meaning of $2 + 3 = 5$ using your cubes? • Why do $3 + 2$ and $2 + 3$ both equal 5? May we explain them using either your cubes or these pictures? • Why did we use subtraction? Can you explain the meaning of your equations? • In small groups, can you explain the meaning of each equation based on the story problem?	**Meaning-making Questions** • What can I do to stay with the same family without changing things? (Point to the cubes) This shows $3 + 2$. • So who can represent $5 - 2$ using your cubes? • If the 3 were the same color, how can you show me?
	Comparison Questions • Compare these two addition equations. What do you find? How are they similar and different? • Compare these two subtraction equations. What do you notice? • Let's compare these four equations only. What do you notice? Why do we call it a fact family? • When do we use addition and when do we use subtraction?	**Comparison Questions** • How does knowing $3 + 2 = 5$ help us with solving these subtraction facts? • Let's compare these three equations, what do you find? (S: They are a fact family!)
Linking to ICAP	Overall, Teacher X's lesson likely prompted students' cognitive engagement up to C and I level. This is evidenced in Teacher X's representation uses and deep questioning as described above.	Overall, Teacher Y's lesson likely prompted students' cognitive engagement up to A level. This is evidenced in Teacher Y's representation uses and questioning (although a few questions have a potential to reach C).
Insights Gained	Answers may vary.	

In particular, regarding the real-world context, Case 1a situated the worked examples ($2 + 3 = 5$, $3 + 2 = 5$, $5 - 2 = 3$, and $5 - 3 = 2$) within a set of inverse word problems about apples. By comparing these problems, students were encouraged to understand the part–whole relationship in a meaningful way. In contrast, Case 1b introduced the apple problems only at the end, emphasizing computation rather than sense-making. When introducing the part–whole model, Teacher Y drew students' attention to the placement of larger and smaller numbers—a procedural focus that is potentially misleading.

3.1.3.3 Deep Questions

In Case 1a, Teacher X asked three types of deep questions: problem-posing questions, meaning-making questions, and comparison questions. These questions focused on fostering meaningful understanding of inverse relationships through real-world contexts. In contrast, in Case 1b, Teacher Y asked only a few deep questions, with a focus on computational answers. This difference is also reflected in how the apple context was approached: in Case 1b, it was used primarily as a pretext for computation.

Problem-Posing Questions. Teacher X began the worked example with a real-world context involving red and green apples. After students described their observations, the teacher asked them to pose a question (e.g., "*Can you ask a question*?"). Students were anticipated to pose two types of questions: (1) "*How many more red apples are there than green apples?*" and (2) "*How many apples are on the table?*" The first question represents a comparison problem to be learned later, while the second aligns with the targeted content and could be adapted by the teacher for instruction. This teaching technique prompts students to generate questions, developing their problem-posing ability—a critical skill highly valued in mathematics education. Using student-generated questions to guide instruction is also likely to foster autonomy and motivate learning. In contrast, this technique was entirely absent in Case 1b, where the set of word problems was simply presented for students to apply learned computation skills.

Meaning-Making Questions. Both Teachers X and Y asked meaning-making questions to promote representational connections. However, the richness and depth of these questions differed. In Case 1a, Teacher X asked students to explain why we used addition (e.g., $2 + 3 = 5$), subtraction ($5 - 2 = 3$), and the commutative property ($2 + 3 = 3 + 2$) using cubes. The teacher also asked students to explain the meaning of number sentences in the context of the story. These questions likely helped students make connections among abstract, semi-abstract, and concrete representations. When Teacher X asked students to explain the meaning of each equation in small groups ("*In small groups, can you explain the meaning of each equation based on the story problem?*"), it further promoted co-construction of knowledge.

Teacher Y also asked questions to encourage connections, requiring students to make inferences. For example, students used cubes to show the related addition fact for $3 + 2$ ("*What can I do to stay with the same family without changing things?*") and to represent $5 - 2$ ("*Who can represent 5 – 2 using your cubes?*"). These questions connected number sentences to cubes but did not link them to the concrete story context. Moreover, the questions focused on factual knowledge rather than the underlying reasoning. Teacher Y could have further encouraged students to explain why or how the cubes represented certain number facts, similar to Teacher X's approach.

Comparison Questions. In Case 1a, Teacher X asked a series of comparison questions, including comparisons between two addition sentences, two subtraction sentences, and all four number sentences in the fact family (see Table 3.2). Students were asked to identify similarities and differences between related number facts and operations (e.g., "*Compare these two addition equations. What do you find? How are they similar and different?*" "*Let's compare these four equations only. What do you notice?*" "*When do we use addition and when do we use subtraction?*"). The anticipated responses were grounded in the part–whole structures. For instance, students were expected to understand that switching the two parts does not change the total ($3 + 2 = 2 + 3$) and that addition is used to find the whole, while subtraction is used to find a part.

In Case 1b, Teacher Y asked only two comparison questions. At the beginning of the lesson, the teacher presented three facts on the board: $3 + 2 =$, $5 - 2 =$, and $5 - 3 =$. After the class discussed $3 + 2 = 5$, the teacher asked, "*How does knowing $3 + 2 = 5$ help us with solving these subtraction facts?*" This question facilitated some connection, but students were only expected to use addition to find the answers for the subtraction facts, rather than to understand the structural relationship between addition and subtraction. Similarly, at the end of the lesson, after solving a set of inverse word problems, the teacher asked, "*Let's compare the three equations. What do you find?*" Students were expected to notice that "they are a fact family," likely meaning only that the three equations involve the same numbers. This can be seen from the beginning of Teacher Y's lesson, where a fact family was defined simply as related facts using the same three numbers. In this sense, the depth of inference-making in Teacher Y's lesson was much shallower than in Teacher X's lesson, where students were required to explain the fact family based on part–whole structures.

3.1.3.4 Linking to ICAP

From the perspective of ICAP, Teacher X's overall lesson likely promotes students' cognitive engagement to at least the *Constructive (C)* level, whereas Teacher Y's lesson is largely limited to the *Active (A)* level. This is evident in the differences in their representation uses and deep questioning described above.

In Case 1a, Teacher X used a concreteness-fading approach, starting the lesson with a concrete context, which provided additional opportunities for deep learning. For example, students were asked to propose questions based on a given picture, going beyond the existing materials. Posing problems or asking questions is a generative behavior that indicates a C level of cognitive engagement (Chi et al., 2018). Teacher X also asked meaning–making questions across various representations. Understanding why the operations and number sentences make sense requires generating new inferences based on what is known, further supporting cognitive engagement at level C. When students explained the meaning of each question in small groups ("*In small groups, can you explain the meaning of each equation based on the story problem?*"), it facilitated co-construction of knowledge, reaching the highest level of cognitive engagement, *Interactive (I)*. Additionally, Teacher X posed a series of comparison questions that prompted students to analyze similarities and differences within and between addition and subtraction facts, as well as inverse relationships. These questions elicited deep responses beyond the provided materials, again likely reaching level C.

In Case 1b, Teacher Y's representation sequence is the reverse of concreteness fading, with the real-world context introduced only at the end as an application of computation. It should be acknowledged that Teacher Y used manipulatives to support discussions of number sentences. However, because the lesson focused heavily on computation, the use of manipulatives may have helped activate prior knowledge but did not generate much new information. Additionally, there were no opportunities for students to pose questions, representing a missed opportunity for deep learning. Although Teacher Y asked some meaning-making and comparison questions, these focused on number facts and computational answers rather than structural understanding. As a result, despite some potential inference-making opportunities, the overall cognitive engagement in Teacher Y's lesson likely remained at level A.

3.1.3.5 Insights Gained

Preservice teachers (PSTs) who have read and discussed this pair of contrasting cases may have different takeaways. Thus, instructors can expect varied responses. While most PSTs may recognize that Teacher X's use of representations and questioning better promotes students' cognitive engagement and learning, some may not agree.

During my first presentation of the contrasting cases in my methods course, one PST shared that she preferred Teacher Y's questions because they seemed easier for students to understand. Another PST felt that both approaches might work, depending on how teachers used them. She explained that she had been taught in a more conventional way, which worked for her, but when other teachers used a different approach, she felt confused. In her course assignment, she further noted that Teacher Y might not be the best representative of a conventional teaching approach because this teacher at least used manipulatives. On the one hand, this was an insightful observation, since many teachers in actual classrooms tend to be more procedural. On the other hand, her comment revealed that some PSTs viewed the mere use of manipulatives as a sign of good teaching. This is precisely why we need to use this contrasting case book: to help PSTs realize that seemingly good math lessons (like Teacher Y's) can still lead to poor learning outcomes (Schoenfeld, 1988).

Nevertheless, our course assignments show that the majority of PSTs learned a great deal from engaging with these contrasting cases (1a and 1b). Many acknowledged that their initial lesson plans resembled those of Teacher Y. After studying this pair of cases, they recognized the value of situating instruction in real-world contexts and then gradually fading from concreteness to abstraction. They also recognized the importance of deep questioning and planned to incorporate it into their future planning and teaching. A typical PST response follows:

> Reflecting on my initial lesson planning approach, I see the value of incorporating real-world contexts and deep questioning into my teaching. Initially, I may have leaned toward focusing on procedural fluency, similar to Teacher Y. However, this case study has highlighted the importance of beginning with concrete examples and encouraging students to explore concepts through guided questioning. In future lessons, I will integrate real-world problems early on and use representations progressively to help students grasp abstract concepts. Additionally, I will prioritize deep questioning to promote critical thinking and interactive discussions

3.2 Teaching Additive Inverses with Comparison Model

3.2.1 Introduction

The Common Core State Standards expect K–2 students to understand the relationship between addition and subtraction. In Case 1, we discussed how this relationship can be taught using part–whole word problems, which can be solved with a fact family. In Case 2, we explore how the relationship can be illustrated through comparison word problems.

STEP 1: THINK AND PREPARE

Imagine you are a second-grade teacher. How would you teach inverse relations (e.g., $3 + 2 = 5$, $2 + 3 = 5$, $5 - 3 = 2$, $5 - 2 = 3$) using comparison word problems? Draft a brief plan in a T–S dialogue format. What representations would you use, and how would you sequence them? What questions, including deep questions, would you ask, and what student explanations would you anticipate?

Your Initial Draft Plan:

3.2.2 Contrasting Cases: 2a and 2b

In Case 2, two subcases illustrate how second-grade teachers approach a worked example of inverse relations (not a full lesson). Case 2a features Teacher X using the TEPS approach, while Case 2b showcases Teacher Y using a conventional method. Both subcases involve a set of word problems based on a comparison model adapted from the Common Core State Standards (see Table 3.3).

Note that students in both classes possess the same prior knowledge, and the two lessons share the same teaching objectives:

Prior knowledge. Students have already learned:

1 Fact families based on the part–whole model
2 "Difference unknown" problems

Teaching objectives. Students will learn:

1 How to find the larger quantity
2 How to find the smaller quantity
3 The inverse relationship between addition and subtraction

STEP 2: READ AND ANNOTATE

Below are Cases 2a and 2b. Read and annotate each case, paying particular attention to the use of representations and questioning. Specifically, for each teacher (X and Y), identify

a The types of representations used
b The deep questions asked

TABLE 3.3 Inverse Word Problems of Additive Comparison Model from the Common Core

Difference Unknown	*Bigger Unknown*	*Smaller Unknown*
Lucy has 3 apples. Julie has 10 apples. How many more apples does Julie have than Lucy?	Julie has 7 more apples than Lucy. Lucy has 3 apples. How many apples does Julie have?	Julie has 7 more apples than Lucy. Julie has 10 apples. How many apples does Lucy have?

Case 2a: Teacher X

1. Review the prior knowledge: Find the difference

(Introduce an image of two children and their names, Lucy and Julie. Then show how many apples each child has.)

T: From the picture, how many apples does each of them have? (Lucy has 3, and Julie has 10).

T: Based on this information, what question might you ask?

S: (Possible responses:)

(1) How many apples do they have in all?
(2) How many more apples does Julie have than Lucy?

T: (Confirm both are correct). We can use counters to represent Lucy's and Julie's apples.
(Ask two students to come to the board and lay out the counters to represent Lucy's and Julie's apples, respectively).

T: Where should we label the question marks for question (1)?

T: Where should we label the question marks for question (2)?
(To label the "?" for the second question, students should line up Lucy's and Julie's apples and separate Julie's apples into two parts: the "same as" part and the "more than" part.) (see Figure 3.11).

T: How would you write an equation to solve each of these problems?

S: (Share solutions):

(1) $10+3=13$
(2) $10-3=7$

T: Why did we use addition to solve question (1) but subtraction to solve question (2)? (let students explain)

T: (Summarize):

- In (1), we are finding the total by putting the two parts together, so we use addition.
- In (2), we are comparing two quantities to find the difference, so we use subtraction.

T: We have learned both types of problems! Now, let's take a closer look at question (2). We asked "how many more." Who can rephrase that question differently?

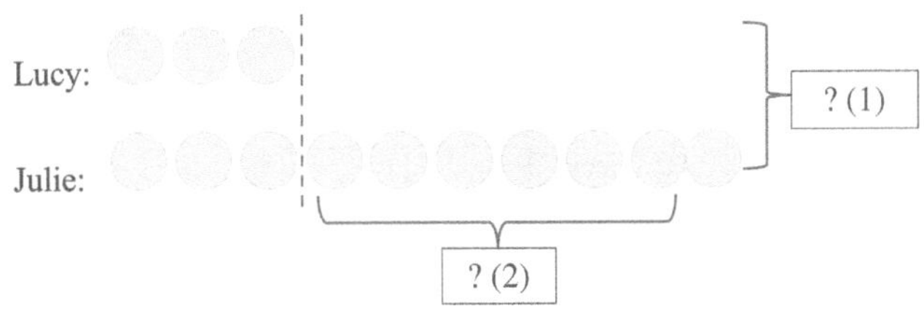

FIGURE 3.11 Board representation for the review task

S: How many fewer apples does Lucy have than Julie?

T: Where in this picture does it show the "how many fewer" part? (Ask a student to come to the board and point it out). How is it related to "how many more" part? (They refer to the same thing).

T: Correct. We call this part the "difference." It doesn't matter if we ask "how many more" or "how many less" — when we find difference, we use subtraction. This is a type of comparison problem.

T: Why do we use subtraction when we find difference? Can someone explain using this picture?

T: (Let students explain in their words. Make sure students see that Julie's apples, or the bigger quantity, consist of two parts: the part that is the same as the smaller quantity, and the difference. If we take away the "same as" part (or the smaller quantity), we will find the difference.

T: (Write on the board): Bigger – Smaller = Difference.

2. Teach how to find the bigger quantity

T: Today, we will continue learning about comparison problems using Lucy and Julie's apples. (Present an example task on the left side of the board.) *Julie has 7 more apples than Lucy. Lucy has 3 apples. How many apples does Julie have?*

T: What does this problem tell us? What do we need to find out? (Let S respond).

T: We can draw a tape diagram to represent each of their apples. If I draw the first tape for Lucy (label it as 3, see Figure 3.12), how can I draw the second tape to represent Julie's apples? Can you please help me? I will use my finger to draw, and you tell me when I should stop. (When it reaches the same length as Tape 1, ask: Can I stop? Why not?)

T: Okay, I am not going to draw it out. You can draw it on your paper. (Ask students to take out the pre-handed paper slip with Tape 1 already drawn. Ask them to complete the second tape to show Julie's apples).

T: (Walk around and select a few different student drawings. Display two drawings: one has a "dashed line" on the second tape while the other does not).

T: I displayed two drawings here (see Figure 3.13). Why did both students draw the 2nd tape longer than the first tape? (Julie has 7 more apples).

T: I noticed that one student added a "dashed line" on the second tape. Why is there a dashed line? What does it tell us? Please discuss this in small groups.

T: (Ask a few students to explain the function of the "dashed line". Students may explain the left and right parts of the line on the second tape). Stress: This line helped us clearly see that Julie's apple consists of how many parts? (2). One the part is the same as Lucy's, which is? (3). The other part is more than Lucy's. How many is it? (7).

T: Based on this diagram, to find out how many apples Julie has, what can we do? (3 + 7 = 10). Why did we add? Can someone use the tape diagram to help us understand this equation? (Let student explain).

FIGURE 3.12 Teacher X drew out the first tape

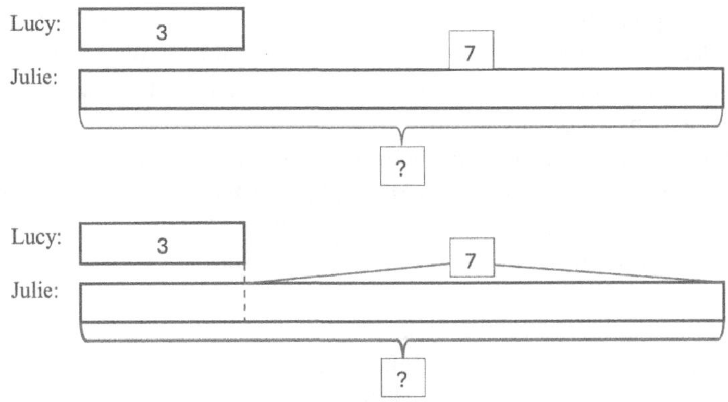

FIGURE 3.13 Two student drawings displayed for comparison

T: (Summarize): The bigger quantity contains two parts: one part is "the same as" part, which is equal to the smaller quantity, and the other is the "more than" part (also called the "difference"). So, "the smaller + difference = the bigger." (write this on the board).

3. Teach how to find the smaller quantity

T: What if I know Julie's apple but not Lucy's? (Present the word problem on the right side of the board): *Julie has 7 more apples than Lucy. Julie has 10 apples. How many apples does Lucy have?*

T: (Ask students to read this problem). What do we know? (Let students explain).

T: On the paper slip, I've drawn one tape to represent Julie's apples. Can you draw the other tape to represent Lucy's? Please complete the diagram on the paper slip.

T: (Display a typical diagram). Can you explain this tape diagram?

T: From the diagram, which two parts do Julie's apples contain? (the same as part, the more than part).

T: To find Lucy's apple, what can we do? (Let students explain. Based on the tape diagram, students should explain that if the "more than part" is taken away from Julie's, what remains is the "same as part," which is equal to Lucy's.).

T: What equation can we use to solve this problem? (10 – 7 = 3). Why did we use subtraction? (Let students explain.)

T: (Summarize:) We take away "the more than part" or "difference" from Julie's apples to obtain "the same as part," which indicates Lucy's apple. So, we use "Bigger – Difference = Small." (write this on the board).

4. Compare the two worked examples

T: We just discussed the two word problems. How are they similar and difference? (Let students explain. Expect them to find out:

- Both problems compare Julie and Lucy's apples.
- One problem asks for the bigger quantity while the other asks for the smaller quantity.
- One problem uses addition to solve while the other used subtraction).

T: In both word problems, there is a word "more". However, we used addition to solve the first problem but subtraction to solve the second problem. Why is that? When do we use addition, and when do we use subtraction? (Remind students we don't solve a word problem solely based on key words).

T: (Summary): The larger quantity contains two parts: the part that is the same as the smaller quantity (also called "the same as part") and the difference (also called "the more than part"). To find the larger quantity, we use: Smaller + Difference = Larger. To find the smaller quantity, we use: Larger – Difference = Smaller.

5. Link the new lesson to the review task

T: Today's lesson builds on what we've already learned. Take a look at our review problem. How are today's example problems different from this review task? (The review task focuses on finding the difference, while today we are either finding the bigger or smaller quantity).

T: This group of word problems is related to each other. They are all about comparisons. When we solve them, we obtain a group of equations (10 – 3 = 7, 7 + 3 = 10, 10 – 7 = 3). We can call them what? (A fact family or related facts).

Case 2b: Teacher Y

1. Review the prior knowledge

(1) **Fact family**

T: Hi friends, we have learned about fact family. Can you use the 3, 7, and 10 to write a fact family? (Let students work on this task individually).

T: Let's share our answers with the class! (Let students share). You all did a great job! What is a fact family? (S: They have the same three numbers.)

T: Remember, we can use our cubes to represent this fact family. Let's look at 10 – 3 = 7. What does it mean? (S: You have 10 cubes, and if you take away 3, you will have 7 left).

T: Correct! Subtraction means take-away. Can we use the cubes to explain the rest number sentences? (3 + 7 = 10, 7 + 3 = 10, 10 – 7 = 3).
(Give students small group time and let them explain the number sentences).

(2) **Find the difference**

T: You all did great! We have also learned about comparison problems. (Take out two cube sticks and project them onto the board)

My red stick has 10 cubes. My green stick has 3 cubes. How many more cubes does my red stick have than my green stick?

T: Can you tell me how many more red cubes I have than green ones? (7)

T: How do you know? Can someone come up to the board to show us why it is 7? (Ask a student to come up to the board, circle the extra part, and count it aloud, see Figure 3.14).

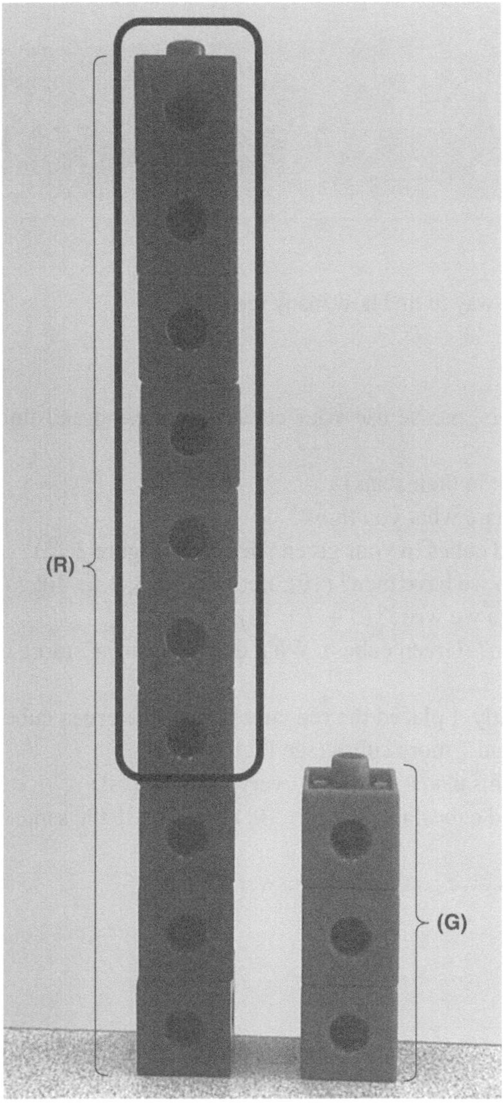

FIGURE 3.14 Two cube sticks for comparison

T: (To the class) Let's count together! (The whole class counts: 1, 2, 3,…7.)
T: It is 7! What equation can we write? (10 – 3 = 7).
T: So, we can subtract the two numbers to find out how many more. Great!

2. Teach how to find the bigger number

T: Today, we will continue learning about comparison problems. (Show the green stick on the board). If I tell you that "*My green stick has 3 cubes and the red sticks has 7 more,*" can you help me figure out, *"How many cubes does my red stick have?"*

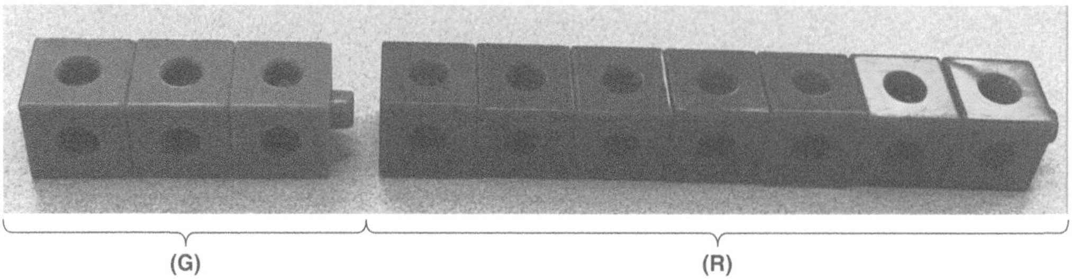

FIGURE 3.15 Student first way to find how many red sticks

T: With your partners, please use your cubes to represent and find how many red cubes there are.
(Students work in their seats)
T: Okay, who can share what you think?
S1: You can add 7 red cubes to your green stick! (see Figure 3.15)
T: Oh, how many do we have then? (10). Let's count. 3, 4, …10!
T: What equation can we write? (3 + 7 = 10).
T: What does 3 mean? (Green cubes). What does 7 mean? (7 more cubes). And 10? (10 red cubes).
S2: I did this differently. I placed the red cubes under the green cubes. So, I put 3 red cubes first and then added 7 more cubes (see Figure 3.16).
T: Great job! Does this also give us 10, everyone? (S: Yes!)
T: Let's count the red cubes! (S: 1, 2, 3…10!) We can use the same equation to represent it, right? (S: Yes!)
T: Great! Both ways give us the same answer, 10!

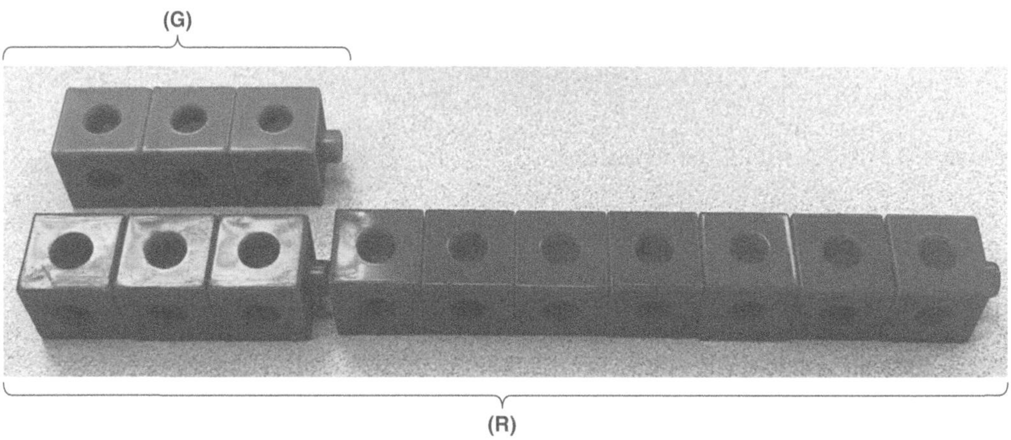

FIGURE 3.16 Student second way to fine how many red sticks

T: We just used addition to find the red cubes, which is a bigger number. In this problem, there is a key word that can help us decide which operation to use. Which keyword is that? (S: More!)
T: Yes! (Circle "more). "The red stick has 7 more," so we use addition!

3. Teach how to find the smaller number

T: What if I only tell you about my red stick this time? (hide green stick). "*My red stick has 10 cubes and the green stick has 7 less.*" Can you help me figure out *"How many cubes does my green stick have?"*
S: Subtraction!
T: What's the answer?
S: 3!
T: How do you know?
S: I remember this from the earlier problem.
T: Great! You made the connection. But what if we don't have the earlier problem? Which equation can we use to solve this problem?
S: $10-7=3$.
T: Why did you use subtraction?
S: I see the keyword "less," so I used subtraction.
T: Awesome! Key words helped us!
T: Let's take out our cubes to check whether this is true. So, we have 10 cubes. We figured out that we have how many green cubes? (3). Okay, is it true that the green stick has 7 less? Count! (S: 1, 2, …7!)
S: Yes!
T: Great job! What did you learn from this example? How should we find something that is "less"?
S: We use subtraction!

4. Wrap up the worked examples

T: We just discussed two new types of comparison problems using the cube sticks. When should we use addition, and when should we use subtraction?
S: When you see "more," we use addition; when we see "less," we use subtraction.
T: Is it easy?
S: Yes!
T: Okay, it's your turn now. Let me give you a few tasks! You can use your cubes to help if needed.
T: (Project the tasks on the board. The lesson then enters the practice section.)

- Lucy has 3 apples. Julie has 10 apples. How many more apples does Julie have than Lucy?
- Julie has 7 more apples than Lucy. Lucy has 3 apples. How many apples does Julie have?
- Julie has 7 more apples than Lucy. Julie has 10 apples. How many apples does Lucy have?

STEP 3: STOP AND REFLECT

Based on your reading and annotating Cases 2a and 2b, please reflect on the following:

1 **Representation uses.** Compare the representation uses of Teachers X and Y. How do their representational sequences differ? How are real-world contexts used differently?
2 **Deep Questioning.** Compare the deep questions asked by Teachers X and Y. How do they differ in terms of frequency and purpose of using deep questions?
3 **Linking to ICAP.** Based on the ICAP framework, which overall level of student engagement does each teacher likely promote? How might each teacher's lesson lead to the respective level?
4 **Insights Gained.** Compare your initial plan with the contrasting cases. What new insights have you gained that you can apply to your lesson planning and future teaching?

3.2.3 Key to Analysis of Case 2

3.2.3.1 An Overview

Cases 2a and 2b share similarities. Both aim to teach the concept of additive inverse relations through a comparison model. Both teachers taught a pair of problems: one is to find the large quantity using addition (7 + 3=10) while the other is to find the small quantity using subtraction (10 – 3=7). Both lessons involved semi-concrete cubes and abstract number sentences.

However, it appears that the lesson foci are different as indicated by teachers' representation uses and questioning. Case 2a focused on sense-making of number sentences and inverse relations through the real-world contexts while case 2b focused on computations. Table 3.4 below listed key points and example deep questions, followed by elaborations.

3.2.3.2 Representation Uses

In Case 2a, the review task started with a real-world context, Lucy and Julie's apples, which was represented by pre-tapes. In the new lesson (worked examples), students were engaged in drawing tape diagrams to understand the quantitative relationships. This leads to solving the problem with abstract equations. Overall, the representational sequence shows concreteness fading. For representational purpose, both the concrete story situations and tape diagrams were used for making sense of the quantitative relationships (e.g., the large quantity = the same as part + the more than parts) and the targeted inverse relationships.

In Case 2b, the review task started with abstract number sentences, which were illustrated by the cubes (an opposite sequence of concreteness fading). In the new lesson (worked examples), it starts with a word problem (e.g., *My red stick has 10 cubes. My green stick has 3 cubes. How many more cubes does my red stick have than my green stick?*), which describes the to-be used cubes. In this sense, the nature of this representation is semi-concrete. The actual real-world context about Lucy and Julie's apples was presented in the end, transitioning to practice tasks. In other words, the real-world contexts were not used for sense-making of the targeted concept. Instead, cubes were used in for new teaching, mainly to find the answers (e.g., counting the "extra" part) rather than to understand the quantitative relationships.

TABLE 3.4 Key Points and Example Deep Questions for Cases 2a and 2b

	Case 2a	*Case 2b*
Representation uses	A real-world context (apples) → tape diagrams → abstract number sentences. This sequence shows *concreteness fading*, and the purpose of real-world context was for sense-making of quantitative relationships.	Abstract number sentences → cubes → word problems This sequence is opposite to concreteness fading, and the purpose of representation uses—including real world context—was for answer-seeking, such as relying on key words.
Deep questioning	**Problem-Posing questions** • Based on this information, what question may you ask? **Meaning-Making questions** • (When draw the tape) Can I stop? Why not? • Why did both students draw the 2nd tape longer than the first tape? • Why is there a dashed line? What does it tell us? • Why did we add? Can someone use the tape diagram to help us understand this equation? • (Display a diagram) Can you explain this tape diagram? **Comparison Questions** • Why did we use addition to solve question (1) but subtraction to solve question (2)? • Why did both students draw the 2nd tape longer than the first tape? • We just discussed the two-word problems. How are they similar and difference? • In both word problems, there is a word "more." However, we used addition to solve the first problem but subtraction to solve the second problem. Why? When do we use addition and when do we use subtraction? • How is today's lesson different from this review task?	**Problem-Posing questions** • NA **Meaning-Making questions** • Can we use the cubes to explain the rest number sentences? • Can someone come up to the board to show us why it is 7? (*but anticipated surface answers*) • Why did you use subtraction? (*but anticipated "key words"*) **Comparison Questions** • When should we use addition and when should we use subtraction? (*but anticipated "key words"*) • *Missed opportunity:* Students' two different cube-stick representations could have been compared; the first representation was incorrect.
Linking to ICAP	Teacher X's instruction potentially promotes' students' cognitive engagement level to at least C. This was evident in Teacher X's use of representations and deep questioning, as described above.	Teacher Y's lesson likely remains at level A. This was evident in Teacher Y's use of representations and questioning as described above.
Insights Gained	Answers may Vary.	

3.2.3.3 Deep Questions

Problem-Posing Questions. Similar to Case 1a, teacher X provided students with one problem-posing question, "*Based on this information, what question may you ask?*" This opportunity is generated by the fact that teacher X started the lesson with a pictorial situation (Julie and Lucy's images each has a few apples). Students were expected to pose two types of questions based on the given information (1) addition (*How many apples do they have in all?*), and (2) subtraction (*How many more apples does Julie have than Lucy?*). Note that question (2) is a comparison problem–finding the difference, a problem type being previously learned. Thus, this problem could serve as a review task for the new lesson. The above opportunity engages students in the process of posing and solving problems, which prompts students till a level C of cognitive engagement (Chi et al., 2018). Unfortunately, problem posing opportunities were completely missing from Case 2b.

Meaning-Making Questions. Both teachers X and Y asked questions to promote meaning-making or representational connections. However, the purpose of questions was quite different.

In Case 2a, tape diagram was uniquely involved in this lesson, generating rich opportunities for deep questioning, which facilitates students' sense-making. For instance, during teacher X's drawing of the tape diagram, this teacher asked students "*Can I stop? Why not?*" This question draws students' attention to the quantitative relationships between Julie and Lucy's apples. Given that Julie has 7 more apples than Lucy, Julie's tape should be longer. Later, after students completed their diagrams on their own, teacher X asked a seemingly trivial but deep question, "Why is there a dashed line? What does it tell us?" Note that comparison-type of word problem is relatively hard for students to grasp. To ensure understanding, it is critical to focus on the relationships among the quantities (the big, the small, the difference). A critical point is that, one needs to grasp the fact that bigger quantity contains (a) the part that is same as the small quantity ("the same as part'), and (b) the part that is more than the small quantity ("more than" part). This shows why we use addition to find the bigger quantity. The above question on the dashed line draws students' attention to this key point. Additionally, teacher X askes students to explain their tape diagrams and use tape diagrams to explain the equations.

In Case 2b, teacher Y asked three questions that potentially facilitate meaning making. The first question—*Can we use the cubes to explain the rest number sentences?*—was raised during the review task. The other two questions were proposed during the new lesson. However, both questions focused only on answers rather than quantitative relationships. For instance, while teacher Y asked, *Can someone come up to the board to show us why it is 7?*, the student only circled the extra part (7) and counted it aloud. This response does not go beyond the existing materials but focused on visual answers. In addition, the teacher asked "*why did you use subtraction?*" which is another potentially deep question. However, the elicited answer is about the key words, "less" means subtraction, which is indeed misleading. Teacher Y's stressing of key words is evidenced on another occasion where she asks, "*In this problem, there is a key word that can help us decide which operation we should use. Which key word is it?*"

Comparison Questions. In Case 1a, teacher X asked a set of comparison questions inviting students to compare two operations (e.g., *Why did we use addition to solve question (1) but subtraction to solve question (2)?*), compare two diagrams (e.g., *Why did both students draw the second tape longer than the first tape?*), compare two worked example tasks (e.g., *We just discussed the two word problems. How are they similar and difference?*), and compare the new and review tasks (e.g., *How is today's lesson different from this review task?).* All of these questions demand students' new inference-making and think beyond the existing materials.

Teacher Y asked one potentially deep question (*When should we use addition and when should we use subtraction?*). Unfortunately, this teacher consistently stressed key words as a response—When you see the key words of "more," we use addition; when we see the key word of "less," we use subtraction. As mentioned above, stressing key words is a strategy often used in classrooms. However, this is not helpful but indeed harmful. This teaching move is different from Teacher X who purposefully argue against key words through a comparison question, "*In both word problems, there is a word* 'more.' *However, we used addition to solve the first problem but subtraction to solve the second problem. Why? When do we use addition and when do we use subtraction?*" Such a question has a potential to facilitate students' understanding of the underlying quantitative relationships, which was summarized by Teacher X at the end of lesson.

In fact, Case 2b contains a couple of missing opportunities for comparison questions. One missed opportunity is that when students provided two different ways to represent the red cubes—the first being incorrect because "green cubes + red cubes ≠ red cubes"—the teacher could have asked them to compare and discuss which representation was correct and why. However, the teacher was satisfied with both ways because both obtained the same answer. The other missing opportunity is that, similar to Teacher X, Teacher Y also taught how to find the bigger and smaller quantities, which could have been compared and contrasted explicitly.

3.2.3.4 Linking to ICAP

From the perspective of ICAP, Teacher X's instruction potentially promotes' students' cognitive engagement level to at least C while Teacher Y's lesson likely remains at level A. This is evidenced from two teachers' representational sequences and purposes. Teacher X started the lesson with the real-world context, which is modeled by the tape diagrams. As seen from the lesson, this teacher spent a significant chunk of time guiding the class to discuss this diagram, focusing on quantitative relationships. These discussions, elicited by types of deep questions (problem-posing, meaning-making, comparisons), were based on, yet went beyond the visible information, prompting students to make new inferences. This indicates C level of cognitive engagement.

In contrast, Teacher Y's lesson mainly limited to cube representations without using real-world context for sense-making. This missed potential opportunities for richer and deeper discussions. As analyzed above, Teacher Y asked a few seemingly deep questions; yet, the anticipated responses either focused on answer-seeking or the keyword strategies. This level of anticipated response does not demand students' reasoning that goes beyond the existing information. As such, it is safe to conclude that Teacher Y's lesson likely remains at level A.

3.2.3.5 Insights Gained

PSTs may gain different insights for planning and teaching from Cases 2a and 2b. Based on my pilot work, I observed that most PSTs expressed interest in—and appreciation for—Teacher X's concreteness-fading approach and the types of deep questioning used, similar to Case 1. They also noted that although Teacher Y employed a hands-on approach, the lesson lacked deep questioning to support meaningful learning. Two typical reflections illustrate this:

> While Teacher Y does use hands-on materials, the focus is more on following patterns and memorizing steps rather than deeply exploring number relationships

> While Teacher Y uses hands-on learning with the cubes, they could have included more in-depth questioning that promotes discussion and critical thinking among peers, like Teacher X did, allowing students to reflect on their thinking and understand why they completed the steps they did

Recognizing that using hands-on materials alone does not guarantee deep learning is an important insight. As the literature shows, many beginning teachers view manipulatives as panacea to student learning, which is superficial (Byrne et al., 2023; Laski et al., 2015). By reading contrasting cases, our PSTs questioned the lack of depth in Teacher Y's use of manipulatives and called for deep questioning, which aligns with the essence of the TEPS approach.

Nevertheless, some PSTs demonstrated inconsistent thinking or lacked critical reflection on the contrasting cases. For example, one PST noted that Teacher Y's lesson prompted interactive engagement because the teacher encouraged students to discuss fact families with their peers. This interpretation is inconsistent with Chi and Boucher (2023), who emphasized that group work alone does not guarantee interactive engagement; without teacher scaffolding, peer discussions may remain at the "active" level rather than achieving deeper cognitive engagement. Similarly, while PSTs in the second quote above perceived that Teacher Y did not ask deep questions, they still believed that both Teacher X's and Teacher Y's lessons offered new insights for them as future teachers. They were particularly impressed when Teacher Y called a student to the board to circle the extra cube of 7, yet they did not recognize that this moment could have been leveraged to teach quantitative relationships rather than merely identifying the correct computational answer.

3.3 Summary

In this chapter, I presented two pairs of cases (1a and 1b; 2a and 2b) to illustrate how one might approach a worked example of additive inverse, using either the TEPS or a conventional approach. The worked example tasks were adapted from the Common Core State Standards, ensuring transferability regardless of textbook choices. In both cases (1a and 2a), Teacher X used the TEPS approach, situating the worked example in a real-world context. This allowed students to pose questions that could later be incorporated into new lessons. The worked examples were then modeled using the concreteness-fading approach, in which manipulatives or drawings were used and further connected to abstract equations. To promote deep learning, Teacher X posed different types of deep questions that helped students make sense of procedures and steps, prompting them to compare across representations, solutions, and tasks. These deep questions encouraged students to make inferences and extend their existing knowledge, often leading to C or I level of cognitive engagement.

In contrast to Teacher X's lessons, Teacher Y in Cases 1b and 2b employed a more conventional approach. At first glance, these lessons appear well-structured; however, they often began with abstract representations without sufficient sense-making and lacked deep questions, such as those involving problem-posing, meaning-making, or comparisons. At times, questions that seemed deep were posed, but Teacher Y's primary goal was to obtain computational answers. Such teaching is likely to result in A or P level of cognitive engagement. Despite these differences, Teacher X's approaches in Cases 1a and 2a offer insights that could help Teacher Y enhance lesson design to better foster students' deep learning.

When implementing the two pairs of contrasting cases, PSTs in my math methods course expressed strong interest not only in the ICAP framework but also in elements of the TEPS approach, such as concreteness fading and deep questioning. While their analyses still have room for growth, they expressed a desire to apply these techniques in their future planning and teaching to support elementary students' mathematical learning.

Note

1 Actual or anticipated student responses appear in parentheses, without necessarily being labeled "S."

References

Byrne, E. M., Jensen, H., Thomsen, B. S., & Ramchandani, P. G. (2023). Educational interventions involving physical manipulatives for improving children's learning and development: A scoping review. *Review of Education, 11*(2), e3400. https://doi.org/10.1002/rev3.3400

Chi, M. T., & Boucher, N. S. (2023). Applying the ICAP framework to improve classroom learning. In C. E. Overson et al. (Eds.). *In their own words: What scholars want you to know about why and how to apply the science of learning in your academic setting* (pp. 94–110). American Psychological Association.

Laski, E. V., Jor'dan, J. R., Daoust, C., & Murray, A. K. (2015). What makes mathematics manipulatives effective? Lessons from cognitive science and Montessori education. *SAGE Open*, *5*(2), 1–8. https://doi.org/10.1177/2158244015589588Maitri Learning+2ResearchGate+2

Schoenfeld, A. H. (1988). When good teaching leads to bad results: The disasters of "well-taught" mathematics courses. *Educational Psychologist*, *23*(2), 145–166. https://doi.org/10.1207/s15326985ep2302_5

4

TEACHING MULTIPLICATIVE INVERSES

Contrasting Cases 3–5

In this chapter, I present three cases (Cases 3, 4, and 5) on teaching multiplicative inverses using different models: equal groups, array, and comparison, respectively. Similar to Chapter 3, each case includes two contrasting sub-cases that address the same mathematical concept. However, teachers' use of representations and deep questions varies considerably, indicating the differences between the TEPS and conventional approaches.

In the TEPS approach (Cases 3a, 4a, and 5a), the worked examples are inverse word problems drawn from the Common Core. Concreteness fading is integrated throughout, often incorporating tape diagrams. Deep questions are systematically posed to create opportunities for problem-posing, meaning-making, and comparison.

By contrast, the conventional approach (Cases 3b, 4b, and 5b) often begins with abstract number sentences modeled with semi-concrete counters or cubes. Real-world contexts may be included, but they are typically introduced at the end, primarily as applications of previously learned computation skills. Teacher questions generally aim to elicit answers with an emphasis on key words. As a result, the meaning of multiplication and quantitative relationships are not emphasized, and comparisons are rarely prompted or made explicit.

Below, we continue to use the three-step process to present the three cases in order to engage readers. As in Chapter 3, we examine the instructional differences and their potential impact on students' learning within the ICAP framework.

4.1 Teaching Multiplicative Inverses with Equal-groups Model

4.1.1 Introduction

The Common Core State Standards expect students in grades 3–5 to understand the relationship between multiplication and division (also called the "*multiplicative inverse*"). This inverse relationship can be illustrated through different problem structures, such as equal groups, arrays, and comparisons. In Case 3, we discuss the teaching of the multiplicative inverse using a set of equal groups word problems drawn from the Common Core.

DOI: 10.4324/9781003586333-4

STEP 1: THINK AND PREPARE

Imagine you are a third-grade teacher. How would you teach inverse relations (e.g., $3 \times 6 = 18$, $18 \div 3 = 6$, $18 \div 6 = 3$) based on an equal groups model? Draft a brief plan in a T–S dialogue format. What representations would you use, and how would you sequence them? What questions, including deep questions, would you ask, and what student explanations would you anticipate?

Your Initial Draft Plan:

4.1.2 Contrasting Cases: 3a and 3b

In Case 3, two subcases illustrate how third-grade teachers approach a worked example of multiplicative inverse relations (rather than a full lesson). Case 3a features Teacher X using the TEPS approach, while Case 3b presents Teacher Y using a conventional method. Both subcases draw on a set of word problems based on equal groups model, provided in the Common Core standards (see Table 4.1).

Note that students in both classes possess the same prior knowledge and the two lessons share the same teaching objectives.

Prior knowledge. Students have already learned:

1 Additive inverse relationships
2 The basic meaning of multiplication, "a groups of b" is represented as a × b

Teaching objectives. Students will learn:

1 How to solve two types of division word problems
2 How division is related to multiplication

STEP 2: READ AND ANNOTATE

Below are Cases 3a and 3b. Read and annotate each case, focusing on the use of representations and questioning. Specifically, for each teacher (X and Y), identify:

a the types of representations used
b the deep questions asked

TABLE 4.1 Equal Groups Inverse Word Problems Adapted from the Common Core

Total Unknown	*Group Size Unknown*	*# of Groups Unknown*
There are 3 bags with 6 plums in each bag. How many plums are there in all?	If 18 plums are shared equally into 3 bags, then how many plums will be in each bag?	If 18 plums are to be packed with 6 in a bag, then how many bags are needed?

Case 3a: Teacher X (TEPS)

1. Review the meaning of multiplication:

T: Friends, do you like fruits? Look at what I have brought for you today! (Show a picture of 3 bags of plums, with 6 plums in each. See Figure 4.1)

T: What do you see? (Let S describe).

T: Okay, there are 3 bags with 6 plums in each bag. Can you ask a question about this? (S: How many plums are there in all?)

T: (Show the complete word problem). *There are 3 bags with 6 plums in each bag. How many plums are there in all?*

T: (Draw a narrow rectangle on the board.) If I draw this part of the tape to represent one bag of plums, can you complete this tape diagram to show what we know and what we need to find out? (Encourages S to work individually, then asks one S to complete the tape on the board. See Figure 4.2).

T: Who can explain this diagram? (Anticipated explanation: The 3 parts represent 3 bags, each part has 6 plums, so we have 3 groups of 6. We need to find the total.)

T: How can you solve this word problem? (S: $3 \times 6 = 18$). Why did we use 3×6? (We have 3 groups of 6. To find the total, we use multiplication).

T: Correct! In this situation, 3 is the number of groups, and 6 is the number in each group, also called the "group size").

T: Someone used 6×3 and got the same answer. Do you agree? Why? (S: 6×3 represents 6 groups of 3, but we have 3 groups of 6. So, we should use 3×6, not 6×3).

T: (Summarize): I agree. When we have equal groups and need to find the total, we can use multiplication, "# of groups × Group size = Total." (Write this on the board.)

FIGURE 4.1 Three bags of plums, with six in each

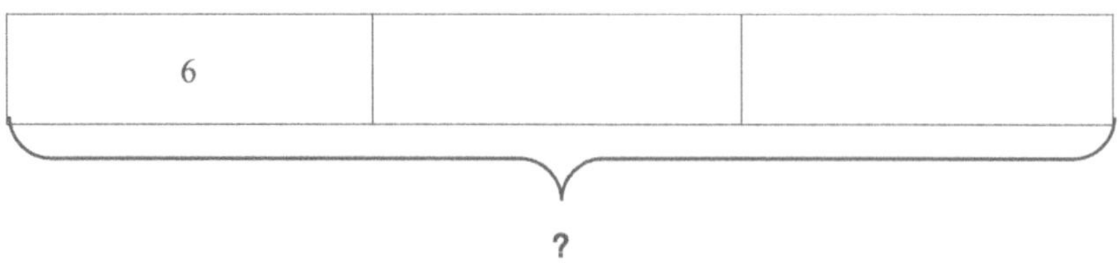

FIGURE 4.2 A tape diagram representing the plum problem situation

2. Teach the first type of division—Find "how many in each group" (group size)

T: Today, we are going to use what we know about multiplication to learn division.

T: Still with the same picture, what if we already know there are a total of 18 plums. (Shows the word problem on the left side of the board:)

If 18 plums are shared equally into 3 bags, then how many plums will be in each bag?

T: In small groups, please take out 18 counters to represent plums. Try to share them equally into 3 bags or 3 groups. How many do you have in each bag?
(Ask a group to come to the board to demonstrate the sharing process.)

T: We can also use a tape diagram to represent this word problem. (Refers to the tape diagram from the review task.) How can we change this diagram to show what we know and what we need to find out? (With students help, following diagram is created. See Figure 4.3).

T: How to find the number of plums in each bag? (Students may answer 6 based on their counter manipulatives). How could we write an equation to solve this? (Students may or may not know).

T: (Based on discussion:) When we know the total and the number of groups—in this task, each bag is one group—and need to find the number in each group (or group size), we can use division. We will write an equation: $18 \div 3 = 6$.

T: What does $18 \div 3 = 6$ mean? Can you explain it based on this plum problem? (Anticipated response: We have 18 plums in total. We shared them equally into 3 bags. We have 6 plums in each bag).

T: In small groups, please explain the meaning of equation again to your deskmate using this tape diagram.

T: (Summarize): When we know the total and the number of groups, and need to find the group size, what operation can we use? (S: Division). So, "Total ÷ # of groups = Group size." (write this on the board).

3. Teach the second type of division—Find "how many groups"

T: I see you have learned this so well. Would you like to accept a small challenge? What if we know the total of plums and that each bag holds 6 plums, but we do not know how many bags? Can you create a word problem for us? You can discuss this in small groups.

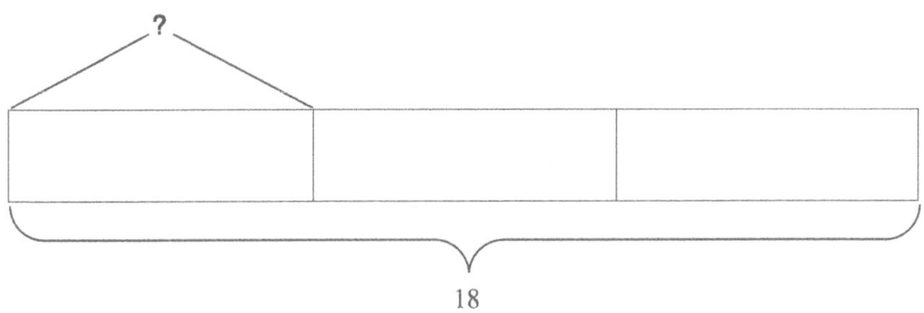

FIGURE 4.3 A tape diagram representing the first division problem

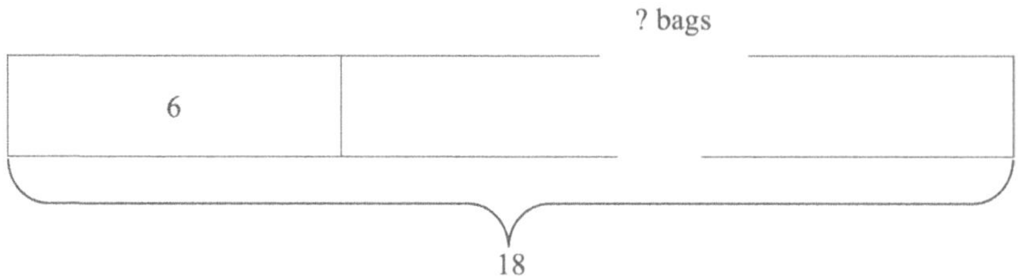

FIGURE 4.4 A tape diagram representing the second division problem

(Ask individual students to share, then post the story problem on the right side of the board).

If 18 plums are to be packed with 6 in a bag, then how many bags are needed?

T: Let's modify the tape diagram. What should we keep? (18). Do we know the number of bags in this task? (No). What do we know? (6 plums per bag). We can keep one bag and erase the other line. We can use a broken tape to show we don't know the number of bags (see Figure 4.4).

T: Can you write an equation to solve this word problem? ($18 \div 6 = 3$).

T: Who can explain the meaning of this equation? (Anticipated response: We have 18 plums in total. We put 6 plums in each bag. So we can pack them into 3 bags).

T: (Summary): For this task, we know the total and the group size, and we need to find # of groups. What do we use? (Division). (T write: Total ÷ Group size = # of groups).

4. Compare two types of divisions

T: Look at these two division problems and their tape diagrams (pointing to the board, see Figure 4.5). What is the same, and what is different? Please compare them with your deskmate.

Students should find:

(1) In both problems, the total is known.
(2) In one problem, we know the number of groups and need to find the group size; in the other, we know the group size and need to find the number of bags.
(3) Division is used to solve both problems.

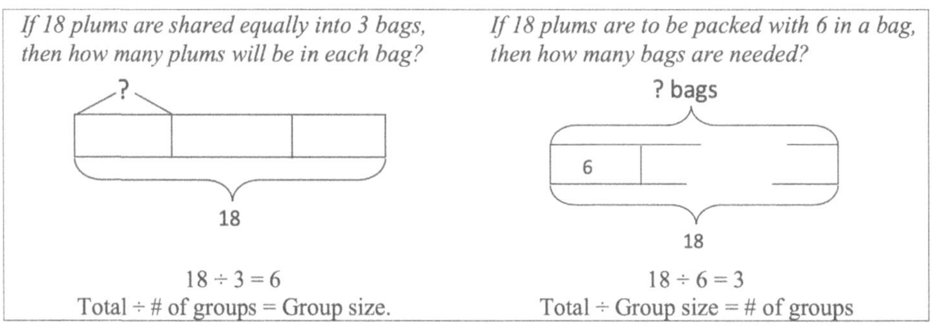

FIGURE 4.5 A comparison of the two division problems

T: Great observations! (Summarize): There are two types of division. One type is used to find the number of groups while the other type is used to find the group size. We use: Total ÷ Group size = # of groups, and Total ÷ # of groups = Group size.

5. Compare division and multiplication

T: How is today's lesson different from the multiplication we learned before? (Refer to the review task). When do we use multiplication and when do we use division?

(S: When we need to find out the total plums, we used multiplication; when we already know the total but need to find number of bags or the number in each bag, we used division).

T: Look at these three equations generated from today's lesson. (Write them in a list: $3 \times 6 = 18$, $18 \div 3 = 6$, $18 \div 6 = 3$). How do they show the relationship between multiplication and division?

(S: They have the same three numbers, with each number representing the same meaning in the story problems)

(T can pointed out that we have the same three quantities. However, multiplication and division are opposite. In multiplication, we need to find the total; In division, we've already known the total.)

T: Because they are opposite, we call multiplication and division "inverse operations." What are the other two inverse operations we learned before? (S: Addition and subtraction).

T: Great! Because they are inverse operations, we can use one to check for the other. For instance, we can use multiplication to check for division, just as we use addition to check subtraction.

Case 3b: Teacher Y (Conventional)

1. Review fact family in addition and subtraction

T: Hi, friends! We have learned about fact family in addition and subtraction. Can you write the other facts in the following fact family? I've written the first one for you:

$2+3=5$

T: Let's check our answers! (Let students report the answers: $3 + 2 = 5$, $5 - 2 = 3$, $5 - 3 = 2$)

T: Correct! How did knowing that $2 + 3 = 5$ help you find the other facts? (They are in a fact family).

T: What is a fact family? (They have the same three numbers). Using one fact in a fact family can help us find the others. Today, we will use the same knowledge to learn about fact family with multiplication and division.

2. Teach fact family in multiplication and division

T: Let's look at this multiplication fact: $3 \times 6 =$ ____

T: What is the answer? (18)

T: How did you find it?

S: $6+6+6=18$.

T: Excellent! Multiplication is repeated addition.

T: Now that we know $3 \times 6 = 18$. Can we write the other facts in the same fact family? Remember, a fact family uses the same three numbers.
S: $6 \times 3 = 18$.
T: How did you get that?
S: I just switched the numbers around.
T: Great! What else? Can we come up with two division sentences? As we just said, a fact family uses the same three numbers.
S: $3 \div 6 = 18$ and $6 \div 3 = 18$
T: Hmm… let's check those. Are these true? Yes, these facts do contain the same three numbers, but let's think about the answers. What is $6 \div 3$?
S: 2!
T: Exactly! $6 \div 3 = 2$, not 18! So, $6 \div 3 = 18$ is incorrect, $3 \div 6 = 18$ is also incorrect.
T: Who else has an idea?
S: $18 \div 3 = 6$ and $18 \div 6 = 3$.
T: Correct! We need to put the biggest number at the front of a division sentence. So, 18 is the biggest. We will use $18 \div 3$ or $18 \div 6$.
T: Let's take a look at this group of facts! What do they have in common?
S: They all have the same three numbers.
T: Great! What are the same three numbers? (3, 6, 18).
T: Does $2 \times 6 = 12$ belong to this fact family? Why not?
S: It does not have the same three numbers.
T: Correct! So, the three numbers —3, 6, and 18— gives us the following fact family.

$3 \times 6 = 18$

$6 \times 3 = 18$

$18 \div 3 = 6$

$18 \div 6 = 3$

3. Illustrate a fact family using cubes

T: We can use cubes to show this fact family.
T: So, what does $3 \times 6 = 18$ mean? (S: It means you have three 6s, like $6 + 6 + 6$).
T: Right! Now, in small groups, please take out three cube sticks, each with 6 cubes. (Wait for students to finish.)
T: How many cube sticks do you have? (3).
T: And, how many cubes are in each stick? (6).
T: So, we have three groups of 6, which is 3×6. How many cubes do we have in total? (18).
T: Now, what if you break each of your cube stick into half? How many cubes are in each smaller stick now? (3).
T: And how many small sticks do you have? (6).
T: Great! So, we now have 6 groups of 3. Do you all agree? (Yes).
T: We can represent it with 6×3. What is the answer? (18).
T: Did we add any cubes? (No). Did we take any cubes away? (No).
T: So, we still have the same number of cubes. 6×3 is also 18.
T: Now, let's say we have 18 cubes. Let's divide into 3 cube sticks of equal length. Please work with your peers! (Wait for students). Are you all done? Count how many cubes are in each cube stick. (6).
T: This tells us that $18 \div 3 = 6$.

T: Now, let's try something else with the 18 cubes. If you put 6 cubes in each stick, how many sticks can you make? We can write this equation as $18 \div 6 = \_\_$.
T: Work with your peers to find out how many sticks you can make. (Wait for students.)
T: How many sticks do you have? (3). Great! So, $18 \div 6 = 3$!
T: We've just used the cubes to illustrate the fact family for multiplication and division. Do you all see how these facts are related? (S: Yes!).

4. Applying fact family to solve word problems

T: We can use our knowledge of fact families to help solve word problems! Let's look at this set of problems about plums. Try to solve them independently. You can refer to this fact family to see which equation might help you.
T: (Project the word problems on the board).

- *There are 3 bags with 6 plums in each bag. How many plums are there in all?*
- *If 18 plums are shared equally into 3 bags, then how many plums will be in each bag?*
- *If 18 plums are to be packed with 6 in a bag, then how many bags are needed?*

T: (After students work independently). Now, check with your peers to see if you have the same answers.
T: Who can share which equation did you choose for the first problem?
S: I used $3 \times 6 = 18$.
T: How do you know?
S: The problem gives me 3 and 6, I can multiply them to get 18.
T: Great job! So, we have a total of 18 plums.
T: Who can share how they solved the second and third problems? Which operations and equations did you use?
S: I used division because we already know 18 is the biggest number. For the second problem, I chose $18 \div 3 = 6$, and for the third problem, I chose $18 \div 3 = 6$.
T: Very good! Now you see how multiplication and division are related — they form a fact family. If we are good at using fact families, we can solve multiplication and division problems more easily.

STEP 3: STOP AND REFLECT

Based on your reading and annotating Cases 3a and 3b, please reflect on the following:

1. **Representation uses.** Compare the representation uses of Teachers X and Y. How do their representational sequences differ? How are real-world contexts used differently?
2. **Deep Questioning.** Compare the deep questions asked by Teachers X and Y. How do they differ in terms of frequency and purpose of using deep questions?
3. **Linking to ICAP.** Based on the ICAP framework, which overall level of student engagement does each teacher likely promote? How might each teacher's lesson lead to the respective level?
4. **Insights Gained.** Compare your initial plan with the contrasting cases. What new insights have you gained that you can apply to your lesson planning and future teaching?

4.1.3 *Key to Analysis of Case 3*

4.1.3.1 *An Overview*

Cases 3a and 3b share several similarities. Both lessons aimed to teach the concept of multiplicative inverse relations through the equal-groups model (illustrated by the related facts $3 \times 6 = 18$, $18 \div 3 = 6$, and $18 \div 6 = 3$, using Common Core story problems). In both cases, the teachers taught a set of related problems: finding the total, finding the number of groups, and finding the number in each group (group size). Both lessons also incorporated story problems, semi-concrete representations (e.g., tape diagrams or cubes), and abstract number sentences—although the sequence and instructional purposes of these representations differed.

Despite these similarities, the lesson foci in Cases 3a and 3b diverged considerably. In Case 3a, Teacher X emphasized sense-making of the multiplication and the two related division word problems, as well as the inverse relationships among them. By contrast, Case 3b centered primarily on computation, with real-world problems serving mainly as applications of computational skills. The purposes for using representations also differed. Specifically, Case 3a employed tape diagrams to highlight quantitative relationships and emphasize the meaning of equal groups. In Case 3b, however, cubes were used largely as directed steps for students to follow, while the set of word problems appeared at the end as an application of previously learned computations.

With regard to deep questions, both the frequency and depth of questioning varied across the two cases. In Case 3a, Teacher X posed questions that probed meaning and quantitative relationships, whereas in Case 3b, questions focused largely on eliciting answers. Table 4.2 below summarizes the key distinctions and provides examples of deep questions, followed by further elaboration.

4.1.3.2 *Representation Uses*

In Case 3a, Teacher X began the lesson with a real-world context: three bags of six plums each. After students described the picture, Teacher X asked the class to pose questions, extending their thinking beyond the given material. The story problem was then modeled with tape diagrams, and students were actively engaged in drawing the tapes while the teacher emphasized the meaning of equal groups. In the new lesson (worked examples), the word problems were modified from the review task using the "variation" technique (Huang & Li, 2017). Tape diagrams were used and compared throughout to help students make sense of the quantitative relationships in division, leading naturally to numerical equations. This representational sequence (concrete → semi-concrete → abstract) reflects *concreteness fading*, with tape diagrams providing rich opportunities for sense-making of quantitative relationships.

In Case 3b, the review task began with abstract number sentences. The worked example also started with abstract fact families, which were then illustrated using cubes. Real-world contexts were introduced afterward, but not for sense-making; rather, they served primarily as applications of learned computation skills. In this sequence (abstract → semi-concrete → concrete), the order is essentially the reverse of concreteness fading, and the use of representations was oriented mainly toward answer-seeking.

4.1.3.3 *Deep Questioning*

Deep questioning in Teachers X and Y's lessons differed across three types: problem-posing, meaning-making, and comparison questions.

TABLE 4.2 Key Points and Example Deep Questions for Cases 3a and 3b

	Case 3a	*Case 3b*
Representation uses	A real-world context (plums) → Tape diagrams→ Number sentences This sequence shows *concreteness fading;* and the purpose of using representations (tape diagrams, real-world context) was to support sense-making of quantitative relationships.	Number sentences→ Cubes→ A real-world context (plums) This above sequence is opposite to concreteness fading; and the purpose of using representations (cubes, real-world context) was for answer-seeking and applying the learned computation.
Deep questioning	**Problem-Posing Questions** • Okay, there are 3 bags with 6 plums in each bag. Can you ask a question? • What if we know the total of plums and that each bag holds 6 plums, but we do not know how many bags? Can you create a word problem for us? **Meaning-Making Questions** • If I draw this part of tape to show one bag of plum, can you complete this tape diagram to show what we've known and what we need to find out? • Who can explain this diagram? • Why did we use 3 × 6? • Someone used 6 × 3 and got the same answer. Do you agree? Why? • How can we change this diagram to show what we've known and what we need to find it out? • What does 18 ÷ 3 = 6 mean? Can you explain it based on this plum problem? • Who can explain the meaning of this equation? **Comparison Questions** • Look at these two division problems and their tape diagrams (board writing). What are the same and what are different? Please compare them with your deskmate. • How is today's lesson different from the multiplication we learned before? (refer to the review task). When do we use multiplication and when do we use division? • Look at these three equations generated from today's lesson. (Write them in a list: 3 × 6 = 18, 18 ÷ 3 = 6, 18 ÷ 6 = 3). How do they show the relationship between multiplication and division? • What are the other two inverse operations that we learned before?	**Problem-Posing Questions** • NA **Meaning-Making Questions** • NA **Comparison Questions** (*but anticipated surface answers*) • Let's take a look at this group of facts! What are common? (S: They all have the same three numbers). • We've just used the cubes to illustrate the fact family for multiplication and division. Do you all see how these facts are related? (S: Yes!).
Link to ICAP	Teacher X's lesson likely promoted students' cognitive engagement to levels C or I, as evidenced by this teacher's use of representations and deep questioning described above.	Teacher Y's lesson likely remains at level A. This is evident in Teacher Y's use of representations and in the absence of deep questioning, given the lesson focus on answer-seeking.
Insights Gained	Answers may vary.	

Problem-Posing Questions. As mentioned above, in Case 3a, Teacher X presented a picture of plums for students to observe. Afterward, students were prompted to pose their own questions: "*Okay, there are 3 bags with 6 plums in each bag. Can you ask a question?*" Later, to transition from the review task to the new lesson, Teacher X again invited students to modify the word problems, providing another problem-posing opportunity: "*What if we know the total number of plums and that each bag holds 6 plums, but we do not know how many bags? Can you create a word problem for us?*" As analyzed in earlier chapters, problem posing is a critical skill that fosters inference-making and, in turn, deep learning. However, such opportunities were completely absent from Teacher Y's lesson in Case 3b.

Meaning-Making Questions. In Case 3a, Teacher X asked a series of meaning-making questions (see Table 4.2). These questions prompted students to either develop a diagram based on the story situation (e.g., *How can we change this diagram to show what we know and what we need to find out*?), explain a diagram (e.g., *Who can explain this diagram*?), or interpret the meaning of a number sentence based on a diagram (e.g., *What does* $18 \div 3 = 6$ *mean? Can you explain it based on this plum problem?*). The purpose of these questions was to encourage representational connections and to help students make sense of abstract ideas.

In Case 3b, although cubes were used to illustrate the number sentences, Teacher Y primarily told students what to do. Students were mainly left to count the resulting cubes rather than make any new inferences. There were no deep questions to promote meaning-making. Additionally, word problems were introduced at the end of the lesson, but the purpose was to show students how knowing number facts can be helpful. Students were not expected to understand the meanings of the number sentences; rather, they were encouraged to rely on surface cues to select an equation. On one occasion, the teacher agreed with students that "We need to put the biggest number in the front of a division sentence," which is mathematically incorrect.

Comparison Questions. In Case 3a, Teacher X asked four comparison questions (see Table 4.2). First, after discussing the two types of division, the teacher asked students to compare them using the tape diagrams: "*What is the same and what is different? Please compare them with your deskmate*." Being an open-ended question, this encouraged students to discuss with peers and co-generate findings. Next, the teacher asked students to compare the new lesson (division) with the review task (multiplication) and to consider: "*When do we use multiplication and when do we use division*?" This was not simply to activate prior knowledge but to support the generation of new inferences, as it explicitly drew students' attention to the conditions for using division, in contrast with its inverse operation. Following these discussions, Teacher X invited students to compare the generated equations and reason about their relationships: "*How do they show the relationship between multiplication and division*?" Students were expected to understand the inverse relationship based on the equal-groups structure, which involves three quantities: group size, number of groups, and total. Finally, Teacher X extended the learning further by asking: "*What are the other two inverse operations that we learned before*?" This question helped bridge connections between multiplicative and additive inverses. This series of comparison questions prompted students to think beyond the existing materials, seek connections between different operations, and develop a structural level of understanding.

In Case 3b, Teacher Y asked two comparison questions; however, the anticipated responses were superficial. The first question asked students to compare the fact family discussed in the new lesson: "*Let's take a look at this group of facts! What do they have in common*?" Students responded, "*They all have the same three numbers*," which the teacher fully accepted. Although this response seems correct, it did not capture the essence of a fact family. In fact, focusing solely on three numbers often leads to mistakes. For instance, as seen in follow-up teaching, students generated two incorrect equations: $3 \div 6 = 18$ and $6 \div 3 = 18$. The second comparison question was: "*We've just used the cubes to illustrate the fact family for multiplication and division. Do you all see how these facts are related?*" Students responded with "*yes*," and Teacher Y appeared satisfied. While students may have seen the relationships, without articulation, their understanding was likely implicit and vague. Teacher Y could have followed up with elaboration on *how* the facts were related to promote deeper learning.

4.1.3.4 Linking to ICAP

From the perspective of ICAP, Teacher X's lesson potentially promotes students' cognitive engagement to the C or I levels, whereas Teacher Y's lesson likely limits students' learning to the A level. This difference is mainly due to the distinct lesson foci, as indicated by the teachers' use of representations and questioning strategies.

In Case 3a, Teacher X's lesson focused on making sense of quantitative relationships and inverse operations. The teacher situated the lesson in a real-world context involving plums, which was modeled using tape diagrams. Students were engaged in drawing and reasoning about these diagrams, making connections between diagrams, story situations, and numerical equations, and comparing types of problems at various levels. They were even encouraged to discuss Teacher X's questions with peers to co-construct knowledge. These processes extended beyond the given materials, promoting deep learning and elevating students' cognitive engagement to the C or I levels.

In contrast, Teacher Y's lesson focused primarily on answer-seeking. Case 3b began immediately with number facts. Cubes were used to illustrate computation results, and story problems applied the learned fact families. Most of the questions posed by Teacher Y focused on computational answers, with responses typically limited to one word. As analyzed above, although there were potentially deep questions, the anticipated answers did not go beyond surface-level understanding (e.g., "They have the same three numbers"). Overall, while this lesson physically engaged students, it did not prompt them to generate new inferences. Consequently, the cognitive engagement in Teacher Y's lesson likely remained at level A.

4.1.3.5 Insights Gained

PSTs may gain different insights for planning and teaching from Cases 3a and 3b. Based on my pilot work, I noticed that while most PSTs favored Teacher X's approaches and concurred with TEPS, some students still identified useful strategies in Teacher Y's approach. For instance, one student wrote:

> I also think that Teacher Y did a great job explaining the concept to the students in some ways. Using some approaches from both teachers will give students a strong understanding of inverse relations in multiplication and division

In some sense, the student was correct, as Teacher Y emphasized computation skills, which were important for promoting computational fluency. However, since computational skills are often emphasized by teachers, the purpose of using contrasting cases is to support deep learning. As such, methods instructors need to make clear although both conceptual and procedural knowledge matter, procedural knowledge by itself does not adequately support elementary students' learning.

Without a solid conceptual foundation, students will struggle to apply their learning in future mathematics.

Nevertheless, I found that PSTs were generally able to differentiate the instructional differences—especially in terms of representation use and deep questioning—between Teachers X and Y, as illustrated by the following typical response:

> During our discussion on how to teach the inverse relationship between multiplication and division, we agreed that we would present each of the two operations with something concrete for the students and gradually introduce increasingly abstract representations. The lesson exhibited by Teacher X was more consistent with our plan and used the concept of concreteness fading when going through multiplication, division, and the inverse relationship between the two. Furthermore, we concluded that we would include more open-ended questions that require deep thinking. While Teacher X asked numerous deep questions, Teacher Y's lesson was more restrictive and relied heavily on one-word-answer questions

The PSTs also demonstrated insights regarding what they could take away from reading Case 3:

> Based on our examination of the two cases, some of the insights we gained include the importance of asking open-ended questions more frequently rather than simple, single-word-answer questions, including various representations that progress from concrete to increasingly abstract, and explaining the connection between operations in a way that students can understand, using language appropriate for teaching this concept

Based on these reflections and takeaways, I consider that the contrasting cases have had a positive influence on PSTs who are learning to teach elementary mathematics in a meaningful way.

4.2 Teaching Multiplicative Inverses with Array Model

4.2.1 Introduction

In Section 4.1, we introduced that the Common Core State Standards' expect students in grades 3–5 to understand the multiplicative inverse, which can be illustrated through different problem structures such as equal groups, arrays, and comparisons. In Case 3, we discussed teaching the multiplicative inverse through a set of word problems based on an equal groups model. In Case 4, we continue this discussion by using an array model, as suggested by the Common Core.

STEP 1: THINK AND PREPARE

Imagine you are a fourth-grade teacher. How would you teach multiplicative inverse (e.g., $3 \times 6 = 18$, $6 \times 3 = 18$, $18 \div 3 = 6$, $18 \div 6 = 3$) based on an array model? Draft a brief plan in a T–S dialogue format. What representations would you use, and how would you sequence them? What questions, including deep questions, would you ask, and what student explanations would you anticipate?

Your Initial Draft Plan:

TABLE 4.3 Inverse Word Problems of Array Model from the Common Core

Total Unknown	*# in Each Row Unknown*	*# of Rows Unknown*
There are 3 rows of apples with 6 apples in each row. How many apples are there?	If 18 apples are arranged into 3 equal rows, how many apples are in each row?	If 18 apples are arranged into equal rows of 6 apples, how many rows will there be?

4.2.2 Contrasting Cases: 4a and 4b

In Case 4, there are two subcases illustrating how two fourth-grade teachers approach a worked example of multiplicative inverse (not a full lesson). Case 4a features Teacher X using the TEPS approach, while Case 4b showcases Teacher Y using a conventional method. Both lessons involved a set of word problems based on an array model, which were suggested by the Common Core (see Table 4.3).

Note that students in both classes possess the same prior knowledge and the two lessons have the same teaching objectives.

Prior Knowledge. Students have learned:

1 The meaning of multiplication, $a \times b$ is denoted as a group of b
2 Multiplicative inverse through an equal groups model

Teaching Objectives. Students will learn:

1 Multiplicative inverse through an array model
2 The commutative property of multiplication through an array model

STEP 2: READ AND ANNOTATE

Below are Cases 4a and 4b. Read and annotate each case, focusing on the use of representations and questioning. Specifically, for each teacher (X and Y), identify:

a the types of representations used
b the deep questions asked

Case 4a: Teacher X (TEPS)

1. Review the prior knowledge: Equal groups

T: (Show a picture of 3 bags of apples, each containing 6 apples. See Figure 4.6). Friends, what have I brought to you today? (Apples!). Each bag has 6 apples. Can we figure out how many apples I brought?

S: 18! $3\times 6=18$.

T: How do you know? Why did you use 3×6? What does 3×6 mean?

T: 3 groups of 6.

T: Someone used 6×3 and got the same answer. Do you agree?

S: No, 6×3 means 6 groups of 3. We only have 3 groups of 6 here.

FIGURE 4.6 Three bags of apples, with 6 in each

2. Teach the array model of multiplication

T: To make them look neat, I decided to rearrange these apples. I want to put each bag in one row, like this (see Figure 4.7).

T: Please take out your counters and use each one to represent an apple. Please help me complete the other two bags of apples.

T: Let take a look at your counters. How many rows of apples do you have now? (3). How many apples does each row have? (6).

T: Here is my question (Present on the board):

There are 3 rows of apples with 6 apples in each row. How many apples are there?

T: If we view each row as one group, can we write an equation to find the total number of apples? (3×6).

T: Why did you use 3×6? What does 3×6 mean?

S: We have 3 rows of 6 apples.

(Introduce the "array" model, draw students' attention to the apple picture on the board).

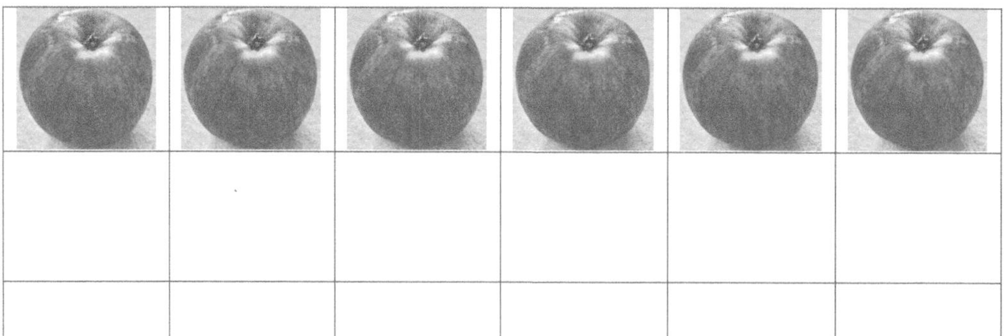

FIGURE 4.7 One bag of apples arranged in an array

FIGURE 4.8 All apples arranged in a 3-by-6 array

T: We just arranged the apples neatly into a format that we call an "array". In this array, we have "rows" and "columns." How many rows do we see? (S: 3). How many columns do we see? (6) (see Figure 4.8).

T: If we view each row as a group (circle the first row), "how many groups of what do we have? (let students think). Yes, we have 3 groups of 6. So, we use $3 \times 6 = 18$ to find out the total.

T: What if I view each column as one group? (Circle the first column). How many groups of what do we have? (6 groups of 3).

T: How do we find the total? ($6 \times 3 = 18$).

T: This is the same as turning this array 90 degrees. (uses a paper picture to demonstrate) In this paper picture, how many rows do we have now? (6). How many are in each row? (3). Correct! This also gives us 6 rows of 3. So, we have $6 \times 3 = 18$ as well.

T: In an array model, we can view one group in different ways, resulting in either 3×6 or 6×3. What does this tell us about the relationship between them? ($3 \times 6 = 6 \times 3$).

T: Does anyone know what this big idea is? (The community property of multiplication).

T: Excellent! Regardless of the arrangement, we still follow the meaning of multiplication because we either see 3 groups of 6 (or 3×6) or 6 groups of 3 (or 6×3) in this array.

3. Teach the array model of division

T: In the past lessons, we learned that multiplication problems can be changed into related division problems. In small groups, can you change the word problem above into two division problems? Recall that in division problems, we must already know what? (total).

T: (give students time to pose two word problems, and walk around to identify different problems and possible mistakes)

T: What is the first word problem you posed? (rephrase students' responses into the task)

If 18 apples are arranged into 3 equal rows, how many apples are in each row?

T: What is the second word problem you posed? (rephrase students' responses into the task).

If 18 apples are arranged into equal rows of 6 apples, how many rows will there be?

T: How should we solve each of these? Please write down an equation on your own notebook. (gives student time).

T: How did you solve these two problems? ($18 \div 3 = 6$; $18 \div 6 = 3$). Explain why you used division to solve each problem. What does each of these equations mean? (Let students explain based on the story problem context)

T: Let's compare these two division problems. How are they similar and different? (Anticipated response: In both problems, the total is known, but one is to find the number of apples in each row while the other is to find out how many rows)

4. Compare multiplication and division & the array and equal groups models

T: Excellent job! Today, we have learned a new model called what? (array). With this model, we can also create multiplication and division problems. Let's take another look at the three word problems. In this model, when do we use multiplication, and when do we use division? (Let students respond and then summarize):

- (We use multiplication to find the total; if we already know the total and need to find either the number of rows or the number in each row, we will use division).

T: Let's look at these pictures (see Figure 4.9). How is the array model related to what we've learned about the equal group problems?

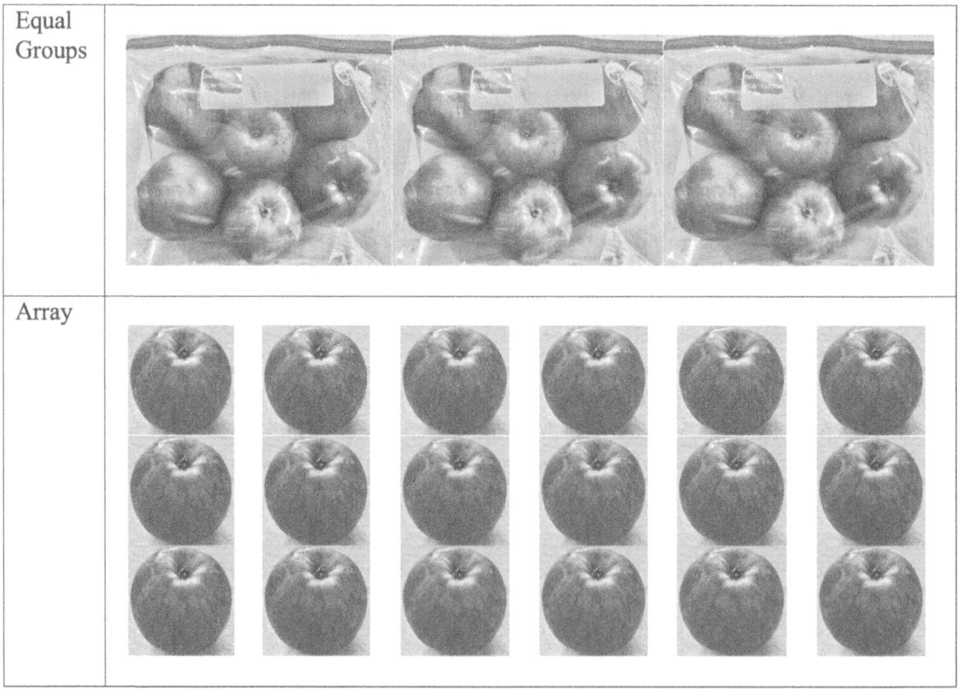

FIGURE 4.9 Apples arranged in equal-groups and array models

S: (Let students discuss in small groups about the similarities and differences between the two models.

T: (Summarize after students' explanation):

- The array model is closely related to the equal groups of model!
- In the equal groups model, the objects in each group are randomly arranged. How many ways can we view one group? (One fixed way: One bag is one group).
- In the array model, the objects are arranged neatly in rows and columns. Therefore, how many ways can we view one group? (Two different ways: Either one row or one column; or you can turn the array around).
- So, in the array model, the two multiplication equations together indicate a big idea called what? (The communicative property).
- Regardless of which model we use, we still follow the basic meaning of multiplication, "a groups of b" is represented as "$a \times b$."
- In both models, if we need to find the total, we will use what? (multiplication); if we already know the total but need to find either of the other factors, we will use? (division).

T: Well done, class!

Case 4b: Teacher Y (Conventional)

1. Review fact family

T: Friends, we have learned about fact family. If I give you three numbers, 2, 3, and 6, what fact family can we come up with? Let's start with the multiplication facts.

S: $2 \times 3 = 6$ and $3 \times 2 = 6$.

T: Great! What about the division facts?

S: $6 \div 2 = 3$ and $6 \div 3 = 2$.

T: Great! It gives us four related facts. What is common in a fact family?

S: They have the same three numbers.

T: Excellent. Today, we will continue learning about fact family, but with an array model.

2. Introduce the term "array"

T: Does anyone know what an array is? An array is a set of objects that are put in rows and columns.

T: (show a piece of grid paper, see Figure 4.10). Do you see the rows and columns? (yes).

T: What is a row? Can someone come up here and show me? (ask a student to come up)

T: What is a column? Can someone show me? (ask another student to come up).

T: Great! We will continue to learn multiplication and division using the array model.

3. Teach multiplication based on an array

T: For this same grid paper, we can use multiplication to find the total number of grids. Let's count. How many grids are in each row?

FIGURE 4.10 An empty array model

S: 6
T: How many grids are in each column?
S: 3
T: We call this a "3-by-6" array. We can use 3×6 to find the total.
T: What is 3×6?
S: 18.
T: How do you know?
S: $6 + 6 + 6 = 18$.
T: Good connection!
T: So, we want to count the rows and columns carefully.
T: I am turning this grid around. Now, what array is this? Which-by-which? Let's count how many in the first column?
S: 6
T: This means we have 6 rows.
T: How many in each row?
S: 3
T: We call this a 6-by-3 array!
T: What multiplication sentence can we write?
S: 6×3
T: What is the answer?
S: 18.
T: Do you see when I turned this array around, I also flipped the numbers? So, $3 \times 6 = 6 \times 3$. We call this the "turned-around property." Mathematicians refer to this as commutative property.

4. Teach division based on an array

T: What if I tell you that there are a total of 18 grids here. Each row has 6 grids (Keeps the first row visible and covers the rest). Do you know how many rows we have?
S: 3!

T: How do you know?

S: $3 \times 6 = 18$, so, $18 \div 6 = 3$.

T: Awesome! You used multiplication to help solve division.

T: Let's see. When we have one row, how many grids do we have? (6). What if we have two rows? (show two rows). How many grids do we have now? (12). (Show one more row). What if we have 3 rows? (18!) Great! So, we have 3 rows.

S: I remember that there were 3 rows in the picture. And, I checked that $6 + 6 + 6$ equals 18. So, there are 3 rows.

T: Good idea. It works too. (points to the array). We have 3 rows.

T: Now, still with 18 grids, what if I tell you that we have 3 rows? (points to the first column and cover the rest) This is the first grid of the 1st row. This is the first grid of the 2nd row. This is the first grid of the 3rd row. So, we have 3 rows. Do you know how many grids are in each row?

S: 6!

T: How do you know? Which equation did you use?

S: $18 \div 3 = 6$

T: How did you figure out that $18 \div 3$ is 6?

S: I know $3 \times 6 = 18$, so $18 \div 3$ is 6.

S: I know $18 \div 6 = 3$, so $18 \div 3 = 6$.

T: All are good ideas! It seems that you see the connections among these number facts!

T: Let's take a look. This array model gives us how many number facts? (4).

T: How many multiplication facts? (2).

T: How many division facts (2).

T: We call them a? (A fact family!)

5. Apply fact family to solve world problems

T: Knowing the fact family with an array model can help us solve real-world problems. See, this is a box of apples arranged in an array model. (see Figure 4.11).

FIGURE 4.11 Apples in an array presented in Teacher Y's class

T: I have three word problems for you. Please solve each of them. Think about the fact family we just discussed; it can help solve these problems. You can discuss with your group members.

1 *There are 3 rows of apples with 6 apples in each row. How many apples are there?*
2 *If 18 apples are arranged into 3 equal rows, how many apples are in each row?*
3 *If 18 apples are arranged into equal rows of 6 apples, how many rows will there be?*

(Let students work in small group.)

T: Okay, let's take a look. Which equation did you choose to solve the first problem?
S: $3 \times 6 = 18$. This word problem tells me "3 rows" and "6 apples in each row," so I choose $3 \times 6 = 18$.
T: Agree? (S: Yes!).
T: Any different ways?
S: $6 \times 3 = 18$. It will give you the same answer.
T: Yes, both will give you the same answer,18 apples.
T: What about the second word problem? Which equation did you choose?
S: I choose $18 \div 3 = 6$. This is because the word problem tells me "18 apples" and "3 equal rows."
T: Great job. What about the third problem?
S: I choose $18 \div 6 = 3$. The word problem tells me "18 apples" and "rows of 6 apples."
T: You are awesome! We can apply our knowledge of fact family to solve related word problems. Well done!

STEP 3: STOP AND REFLECT

Based on your reading and annotating Cases 4a and 4b, please reflect on the following:

1 **Representation uses.** Compare the representation uses of Teachers X and Y. How do their representational sequences differ? How are real-world contexts used differently?
2 **Deep Questioning.** Compare the deep questions asked by Teachers X and Y. How do they differ in terms of frequency and purpose of using deep questions?
3 **Linking to ICAP.** Based on the ICAP framework, which overall level of student engagement does each teacher likely promote? How might each teacher's lesson lead to the respective level?
4 **Insights Gained.** Compare your initial plan with the contrasting cases. What new insights have you gained that you can apply to your lesson planning and future teaching?

4.2.3 Key to Analysis of Case 4

4.2.3.1 An Overview

The two lessons in Case 4 share similarities. Both aimed to teach the concept of the multiplicative inverse through an array model. In particular, they addressed the related facts: $3 \times 6 = 18$, $6 \times 3 = 18$, $18 \div 3 = 6$, and $18 \div 6 = 3$, using Common Core suggested story problems. These word problems involved finding the total, finding the number of rows, and finding the number in each row, respectively. Additionally, both lessons incorporated story problems, semi-concrete representations (e.g., arrays), and abstract number sentences.

However, the lesson foci differed. Case 4a emphasized sense-making of the multiplication and division word problems as well as the inverse relationship between them. In contrast, Case 4b emphasized computations, with the real-world problems serving primarily as applications of computational skills. These differences were evident in the use of representations (sequence and real-world connection) and in teachers' questioning (frequency and purpose). Table 4.4 below lists the key points and example deep questions, followed by elaborations.

4.2.3.2 Representation Uses

In Case 4a, Teacher X began the review task with a real-world context—bags of apples—using an equal groups model. When discussing the number sentences, Teacher X emphasized the basic meaning of multiplication, laying a foundation for subsequent reasoning. In the new lesson (worked examples), the bags of apples (equal groups model) were rearranged into an array model. After discussing the multiplication problem, students were encouraged to transform it into two division problems and solve them. The representational sequence reflected *concreteness fading*. Building on the generated representations (story problems, array, number sentences), comparisons were made among the related word problems as well as across different types of models (comparison and equal groups). This indicates that the purpose of representations was to promote sense-making of quantitative relationships and structures.

In Case 4b, Teacher Y began the review task with abstract number sentences (a fact family). In the new lesson (worked example), a definition of "array" was introduced first. Next, multiplication and related division were taught using the array model but without contextual support. However, discussion of the array model centered on computation—that is, ways to obtain the answer. Finally, real-world contexts (a set of Common Core word problems) were presented, functioning primarily as a pretext to apply the learned computational skills. No discussions about the quantitative relationships were observed. In this sense, the representational sequence (abstract → semi-concrete → concrete) was the reverse of concreteness fading, with the main purpose being answer-seeking.

4.2.3.3 Deep Questions

Similar to the earlier cases, in Case 4, deep questioning in Teachers X and Y's lessons differed across three types of questions: problem-posing, meaning-making, and comparison.

Problem-Posing Questions. In Case 4a, Teacher X provided one opportunity for students to pose questions: "In small groups, can you change the above word problem to two division problems?" Changing a multiplication problem into corresponding division problems was neither simple nor easy. Thus, Teacher X encouraged students to work in small groups. As seen in Cases 1–3, asking problem-posing questions is a common practice in the TEPS approach, which facilitates deep learning. Perhaps due to the fully packed content of this lesson, Teacher X did not ask students to pose additional questions, demonstrating flexibility in using this questioning technique. In contrast, in Case 4b, Teacher Y directly presented the word problems to students, missing opportunities for problem-posing.

Meaning-Making Questions. In Case 4a, Teacher X asked a set of meaning-making questions, especially focusing on the meaning of multiplication. Research (Ding & Heffernan, 2018)

TABLE 4.4 Key Points and Example Deep Questions for Cases 4a and 4b

	Case 4a	*Case 4b*
Representation uses	A real-world context→ Array→ Number sentences. This sequence shows *concreteness fading;* and the purpose of using representations, including the real-world context, was to support sense-making of quantitative relationships.	Number sentences→ Array→ A real-world context. This sequence (abstract, semi-concrete, concrete) is opposite to concreteness fading; and the purpose of using representations, including the real-word context, was primarily for answer-seeking.
Deep questioning	**Problem-Posing Questions** • In small groups, can you change the above word problem to two division problems? **Meaning-Making Questions** • Why did you use 3 × 6? What does 3 × 6 mean? (3 groups of 6). • Someone used 6 × 3 and got the same answer. Do you agree? (No, 6 × 3 means 6 groups of 3. We only have 3 groups of 6 here). • Why did you use 3 × 6? What does 3 × 6 mean? (We have 3 rows of 6 apples). • Explain why you used division to solve each problem. What does each of these equations mean? **Comparison Questions** • In an array model, we can view one group in different ways, resulting in either 3 × 6 or 6 × 3. What does this tell us about the relationship between them? • Let's compare these two division problems. How are they similar and different? • Let's take another look at the three word problems. In this model, when do we use multiplication and when do we use division? • How is the array model related to what we learned about the equal group meaning?	**Problem-Posing Questions** • NA **Meaning-Making Questions** • NA **Comparison Questions** • What is common in a fact family? (*but anticipated surface level of responses, "They have the same three numbers"*).
Linking to ICAP	Likely promoted students' cognitive engagement up to C or I level. This is evident in Teacher X's teaching focus on quantitative reasoning, as indicated by the teachers' use of representation and deep questioning.	Likely prompted students' cognitive engagement up to A-level. This is evident in Teacher Y's focus on computational answers, as indicated by this teacher's representation uses and the absence of deep questioning.
Insights Gained	Answers may vary.	

shows that students—and even teachers—often have a blurred understanding of the basic meaning of multiplication. For instance, based on the U.S. convention, 3 × 6 should refer to 3 groups of 6 rather than 6 groups of 3. However, many interpret 3 × 6 either way, a misconception that can hinder mathematical reasoning.

During the review in Case 4a, the teacher asked, "*Why did you use 3 × 6? What does 3 × 6 mean?*" Teacher X also highlighted the misconception to students' attention: "*Someone used 6 × 3 and got the same answer. Do you agree?*" This discussion emphasized meaning-making and laid a foundation for later teaching using the array model. During the new lesson, Teacher X continued to ask, "*Why did you use 3 × 6? What does 3 × 6 mean?*" Students were also asked to explain why they used division to solve each problem and what each equation represented.

In contrast, in Case 4b, Teacher Y focused entirely on answer-seeking. For example, the meaning of "*3 × 6*" was not discussed. Instead, Teacher Y told students, "*We call this a '3-by-6' array. We can use 3 × 6 to find the total.*" When students solved the division problem, the teacher asked, "*Which equation did you use?*" Students responded with *18 ÷ 3 = 6*. The teacher then asked, "*How did you figure out 18 ÷ 3 is 6?*" The above set of the questions primarily focused on computation rather than sense-making. At the end of the lesson, when discussing the word problems, the teachers asked students which equations they had chosen. The students' explanations appeared to rely on the given individual quantities (e.g., 18 apples, 3 equal rows) rather than the relationships between them.

Comparison Questions. In Case 4a, Teacher X asked students to make various comparisons, including comparisons between solutions, operations, and different types of models. After the class viewed the array model in two ways to generate two expressions (3 × 6 or 6 × 3), Teacher X asked, "*What does this tell us about the relationship between them?*"—eliciting the big idea of the commutative property. Later, when the class solved two division problems, Teacher X asked, "*Let's compare these two division problems. How are they similar and different?*" This drew students' attention to the structural connection between the two types of division. Next, the teacher asked students to compare different operations: "*Let's take another look at the three word problems. In this model, when do we use multiplication and when do we use division?*" This provided an opportunity to understand inverse relations. At the end of the lesson, the teacher also asked, "*How is the array model related to what we learned about the equal group meaning*?" To assist this comparison, pictures were presented. This likely furthered students' understanding of the new concept—the array model—and how it was built upon the equal groups model. All these comparison questions prompted students to go beyond the existing materials or simply generating numerical solutions, and to understand the structural connections.

In Case 4b, Teacher Y asked only one comparison question during the review task: "*What is common in a fact family*?" This question had the potential to facilitate deep learning. However, students were only expected to identify that a fact family involves the same three numbers. As in earlier cases, such responses limited to the given materials and remain at a surface level. Moreover, there were missed opportunities to ask comparison questions. For instance, the array was used to generate two multiplication facts, but no comparison was made to discuss the underlying big idea. Instead, Teacher Y simply stated, "*Do you see when I turned this array around, I also flipped the numbers around? So, 3 × 6 = 6 × 3. We call this a turned-around property.*" Another example is that Teacher Y provided a group of related word problems at the end, linked to the fact family; however, no comparisons were made to help students understand the quantitative relationships and the multiplicative inverse.

4.2.3.4 Linking to ICAP

From the perspective of ICAP, the lesson in Case 4a likely promoted students' cognitive engagement up to level C or I. In contrast, the lesson in Case 4b may have engaged students only at level A. This difference is evidenced by Teachers X and Y's distinct use of representations and questioning, as analyzed above. For instance, in Case 4a, Teacher X employed *concreteness fading* with the array model, providing opportunities for inference-making. The three types of unique questions—problem-posing, meaning-making, and comparisons—further encouraged students to think beyond the given materials. In Case 4b, these opportunities for deep questioning were almost entirely absent. Representations were used solely to support answer-seeking, and students were not prompted to extend their thinking beyond the provided materials, likely keeping their engagement at level A.

4.2.3.5 Insights Gained

By reading Cases 4a and 4b, PSTs may gain different insights for planning and teaching. Based on my pilot work, PSTs can identify the strengths in representation use and deep questioning in Teacher X's approaches when compared with Teacher Y's. Below is a typical response that indicates PSTs' learning based on the contrasting cases, linked to the ICAP model:

> After looking at both teachers' approaches to the array lesson, I immediately noticed a difference in how deeply students were engaged with the material. Teacher X's lesson did a better job following the ICAP model. In this lesson, Teacher X fostered interactive and constructive learning. For example, when teaching the lesson, Teacher X said, 'In small groups, can you change the word problem above into two division problems? Recall that in division problems, we must already know what?' This prompts students to work collaboratively while building off each other's knowledge to come up with their own questions on division. In contrast, Teacher Y's lesson leans more toward active and passive engagement. There is nothing wrong with this—fact fluency and structured explanations have their place in math instruction—but it lacks the depth of engagement that Teacher X's lesson provides. This case made me recognize that a balance between both teaching strategies is necessary.

In the above response, the PST rightly analyzed teacher X and Y's instructional differences and highlighted that Teacher Y's lesson lacked depth. However, this PST still thought that a balance between Teachers X and Y was necessary. While an equal emphasis on procedural and conceptual knowledge is important, given that the contrasting cases focused only on the worked-example portion (new teaching) of a lesson, I would like to emphasize again the conceptual focus (Teacher X's lesson) rather than the computational one (Teacher Y's lesson). Methods instructors may draw PSTs' attention to this emphasis.

In addition to deep questioning, PSTs also gained insights about the use of representations. A typical response illustrates this:

> By viewing the equation alongside the visual, students will understand that multiplication is the same as counting groups (rows and columns), as well as reinforcing the idea of repeated addition. We will definitely apply this new insight to current and future lesson planning, as it provides both concrete and abstract clarification and representation, fostering deeper understanding

In this response, the PST highlighted the importance of visual support for understanding abstract equations and the meaning of multiplication. Across PSTs' responses, they recognized how representations could support meaning-making alongside deep questioning. Many PSTs indicated tha they planned to apply these insights in their future planning and teaching to promote students' deep understanding.

4.3 Teaching Multiplicative Inverses with Comparison Model

4.3.1 *Introduction*

The Common Core State Standards expect students in grades 3–5 to understand the relationship between multiplication and division. This inverse relationship can be illustrated through different problem structures, including equal groups, arrays, and comparison problems. In Case 5, we discuss teaching the multiplicative inverse relationship using a set of comparison word problems provided by the Common Core. Note that comparison word problems are the most difficult type for students because they involve relationships that are hard to manipulate (Riley & Greeno, 1988; Stern, 1993). Thus, key concepts like "3 times" demand careful instructional design to develop students' meaningful understanding.

STEP 1: THINK AND PREPARE

Imagine you are a fourth-grade teacher. How would you teach multiplicative inverse (e.g., $3 \times 6 = 18$, $18 \div 3 = 6$, $18 \div 6 = 3$) based on a comparison model? Draft a brief plan in a T–S dialogue format. What representations would you use, and how would you sequence them? What questions, including deep questions, would you ask, and what student explanations would you anticipate?

Your Initial Draft Plan:

4.3.2 *Contrasting Casers: 5a and 5b*

In Case 5, there are two subcases illustrating how two fourth-grade teachers approach a worked example of inverse relations (not a full lesson). Case 5a features Teacher X using the TEPS approach, while Case 5b showcases Teacher Y using a conventional method. Both lessons involved a set of word problems based on comparison model suggested by the Common Core standards (see Table 4.5).

TABLE 4.5 Inverse Word Problems of Multiplicative Comparison Model from the Common Core

Multiple Unknown	*Bigger Unknown*	*Smaller Unknown*
A red hat costs $18 and a blue hat costs $6. How many times does the red hat cost as much as the blue hat?	*A blue hat costs $6. A red hat costs 3 times as much as the blue hat. How much does the red hat cost?*	*A red hat costs $18 and that is 3 times as much as a blue hat costs. How much does a blue hat cost?*

Note that students in both classes possess the same prior knowledge and the two lessons have the same teaching objectives.

Students' prior knowledge. Students have already learned:

1 The meaning of multiplication & division based on the equal groups model
2 Multiplicative inverses using the equal group models
3 One type of comparison problem, multiple unknown (find "how many times")

Teaching objectives. Students will learn:

1 The other two types of comparison problems (finding the bigger, finding the smaller)
2 How multiplication and division are related in comparison word problems

STEP 2: READ AND ANNOTATE

Below are Cases 5a and 5b. Read and annotate each case, focusing on the use of representations and questioning. Specifically, for each teacher (X and Y), identify:

a the types of representations used
b the deep questions asked

Case 5a: Teacher X (TEPS)

1. Review the prior knowledge

(1) **The equal groups model**

T: Friends! We have been learning about multiplication and division recently, and you've done a great job! Let's see what problem I brought for you today! (Show a picture of 3 hats, each priced at $6). A store is selling this type of hat (see Figure 4.12). Who can create a multiplication problem based on this?

T: (Asks a few students to share.) Great! Mine is similar to what you just came up with. (Show the problem on the board.)

- *Tom bought 3 hats from a store, each costing $6. How much did Tom spend in all?*

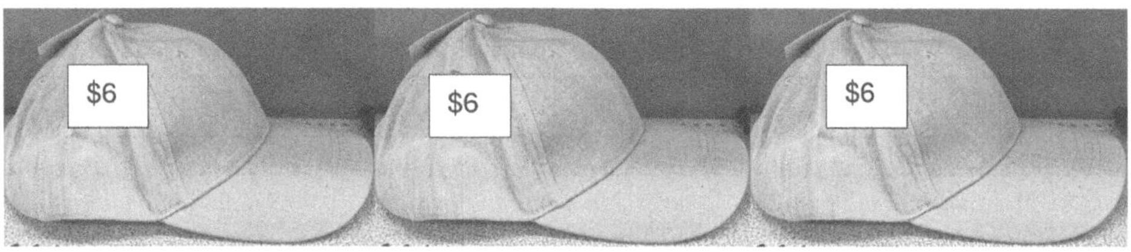

FIGURE 4.12 Three hats, $6 each

T: How do we solve this? ($3 \times 6 = 18$). Why did we use 3×6? (3 groups of $6). Great! To find the total for equal groups, we use multiplication.

T: Now, what if we know that Tom spent $18? Can you change this problem into two division problems? (Possible answers:)

- *Tom spent $18 to buy hats. Each hat costs $6. How many hats did Tom buy?*
- *Tom spent $18 to buy 3 hats from a store. How much does each hat cost?*

T: Please solve these two word problems on your own. (Let students share their solutions in class). Who can explain the meaning of your equations?

T: Let's compare these word problems. What is similar, and what is different? When do we use multiplication, and when do we use division? (Invite students to explain.)

T: (Based on the student responses, summarize on the board:)

- Number of groups × Group size = Total
- Total ÷ number of groups = Group size
- Total ÷ Group size = Number of groups

(2) **Basic comparison problem: To find the multiple**

T: In addition to the blue hat that costs $6, this store also sells the red hat. (Show the picture in Figure 4.13) It is more expensive! If the red hat costs $18. How many times does the red hat cost as much as the blue hat?

T: We have already learned this type of problem. Can you write an equation to solve it?

T: (Gives students time to solve independently. Students should come up with $18 \div 6 = 3$).

T: Can you explain why you used division? (We need to find how many times the red hat costs as much as the blue hat. In other words, we need to find out how many times $6 goes into $18. So, we use division. It is 3 times).

T: Okay, can someone explain to us what "3 times" actually means? (It means 3 groups of the same amount). Yes, it means there are 3 groups of the same amount as what we are comparing.

2. Teach new comparison problem: To find the bigger quantity

T: Today, we will continue learning about comparison problems. (Show on the left side of the board):

A blue hat costs $6. A red hat costs 3 times as much as the blue hat. How much does the red hat cost?

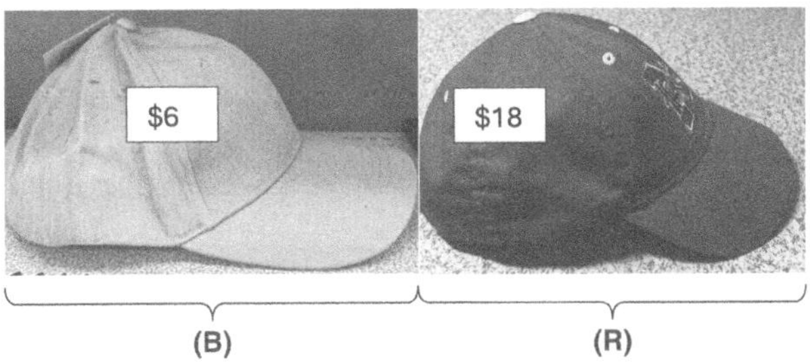

FIGURE 4.13 Blue and red hats with different prices

T: What do we already know, and what do we need to find out? (S: We know the price of the blue hat. We know the red hat costs 3 times as much. We need to find out the cost of the red hat).

T: If I use one chip to represent $1, how many chips will I need to represent the costs of the blue hat? (6).

T: Great! (Display 6 chips).

T: We know that the cost of the red hat is 3 times as much as the blue hat. How can we use the chips to show the cost for the red hat? (Students will likely come up with the picture shown in Figure 4.14a).

T: Who can help me quickly see that the cost of the red hat is "3 times" as much as the blue? (Ask a student to come to the board to adjust the chips, leaving space between groups. See Figure 4.14b).

T: Now, can you explain how this shows "3 times"? (S: If we view the blue as one group, we see that the red contains 3 of these groups. So, we say the red hat costs 3 times as much as the blue hat.)

T: Great! If we box each group, it will show "3 times" even more clearly. (see Figure 4.14c). We can see that the cost of the blue hat has one unit, while the cost of the red hat contains how many units? (S: 3 units, each equal to the blue hat.)

T: If we remove the chips, it will leave us with a tape diagram like this (see Figure 4.14d):

T: Who can explain this tape diagram based on the hat problem? (Ask students to explain.)

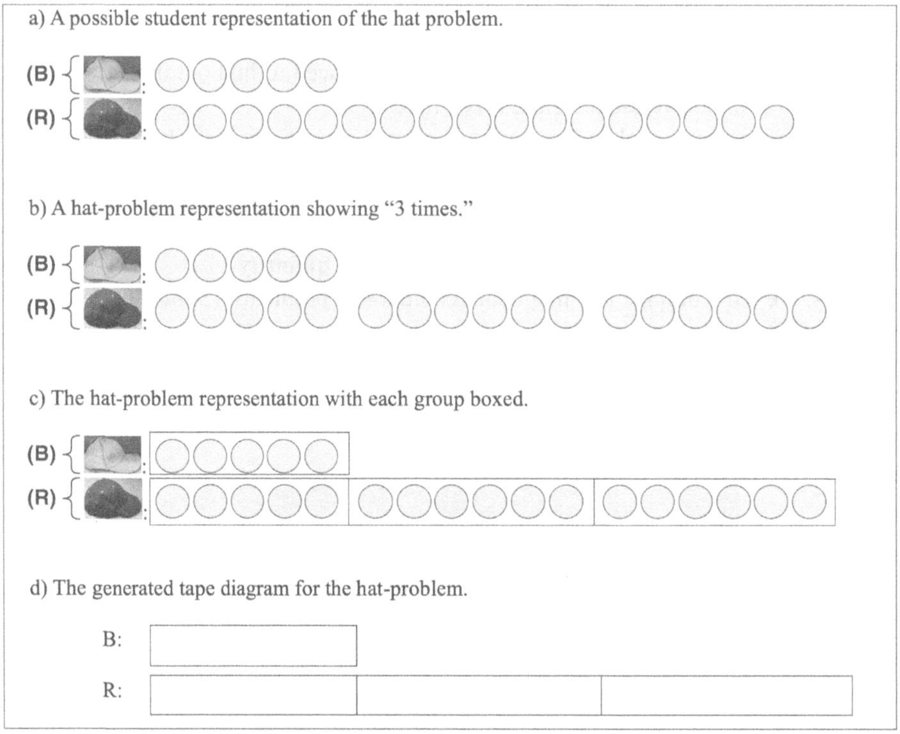

FIGURE 4.14 Transitioning from chips to a tape diagram

T: Why did we draw 3 sections in the second tape? (S: Because it says "3 times")
T: What does "3 times" mean? (S: If we view the cost of the blue hat as one group or one unit, the cost of the red hat has 3 groups of the same amount as the blue hat).
T: (Erase the little dashes). What do you think about this tape? (S: It does not show 3 times).
T: (Add back the little dashes). Correct! "3 times" means that the red hat has 3 groups of the same cost as the blue hat, which is 3 groups of $6.
T: How can we write an equation to find the cost of the blue hat? ($3 \times 6 = 18$). Why do we use multiplication? (The red hat has "3 groups of $6," so, we use multiplication).
T: Excellent! In this word problem, the blue hat is a smaller quantity, and the red hat is a bigger quantity. The bigger quantity contains 3 groups of the same amount as the smaller quantity. Therefore, to find the bigger quantity, we use multiplication.

3. Teach new comparison problem: To find the smaller quantity

T: What if we know the cost of a red hat, but we need to find the cost of the blue hat? (Show on the right side of the board):

A red hat costs $18 and that is 3 times as much as a blue hat costs. How much does a blue hat cost?

T: What do we already know, and what do we need to find out?
T: Let me draw the cost of the red hat (see Figure 4.15).
T: How can we draw the second tape to represent the blue hat? (Ask students to discuss in small groups).
T: (Hand out a pre-drawn tape and let students complete the second tape. To correctly draw the tape for the blue hat, Students need to first break the red hat tape into 3 equal groups).
T: (If students directly draw a short tape for the blue hat without breaking the first tape, then ask:) How do you know the tape for the blue hat should be this length? What does this problem tell us about the relationship between the red and blue hats? (S: 3 times).
T: How can we show "3 times" on the tape? What does "3 times" mean? (S: The red hat tape contains 3 groups or units of the same amount as the blue. So, we can break this tape into 3 equal groups). (T breaks the first tape into 3).
T: How will this help us draw the tape for the blue hat? (S: It has one unit or one group). (T labels the second tape. See Figure 4.16).
T: According to this diagram, how can we find the cost for the blue hat? ($18 \div 3 = 6$).
T: Why do we use division? Can you explain your equation based on the diagram? (S: When we divide the cost of the red hat, $18, into 3 equal groups, each group is $6. That's the same as the cost of the blue hat).
T: Great! In this problem, we need to find the smaller quantity, so we use division.

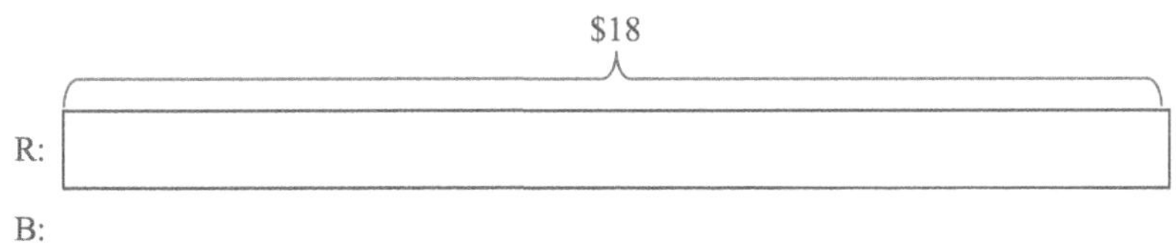

FIGURE 4.15 Tape diagram for the division problem, with the red hat drawn

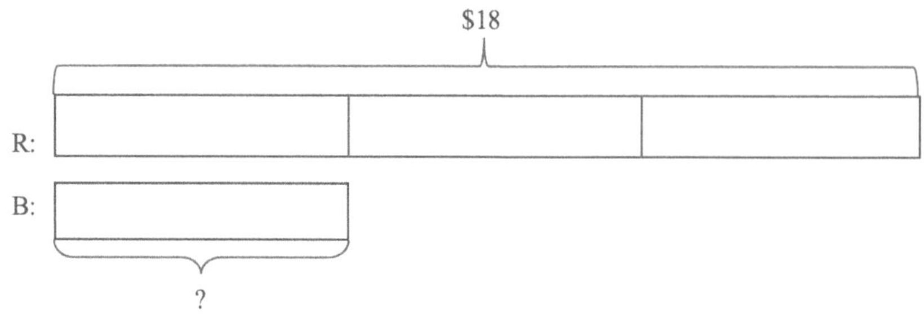

FIGURE 4.16 A complete tape diagram for the division problem

4. Compare the two types of comparison problems

T: Let's take a look at these two problems. How are they similar, and how are they different? (Let students discuss and explain. T then highlight:)

- They are both about comparing the costs of the red and blue hats.
- In both problems, the relationship between their costs is "3 times."
- One is to find the cost of the red hat, and the other is to find the cost of the blue hat.
- One uses multiplication, and the other uses division.

T: When do we use multiplication, and when do we use division? (S: To find the bigger quantity, we use multiplication; to find the smaller quantity, we use division.)

5. Compare the whole set of comparison problems

T: Today, we learned two new types of comparison word problems. We either find the bigger or smaller quantity. How are today's example problems similar to, but different from, our review task here?

(*S:* In the review problem, we know both the bigger and smaller quantities, and we need to find out how many times as much one is as the other.)

T: That's right! We call this relationship "multiple." To find the multiple, we use? (Division).

T: Across these three problems, we have the same three quantities, the bigger, the smaller, and the multiple. Each time, we know two of these and need to find the third. To find the multiple, we use? (Division). To find the bigger, we use? (Multiplication). To find the smaller, we use? (Division).

T: (Write on the board):

Bigger ÷ Smaller = Multiple

Multiple × Smaller = Bigger

Bigger ÷ Multiple = Smaller

6. Compare the "comparison" and the "equal groups" problems

T: At the beginning of today's lesson, we reviewed equal groups word problems. (Show the "3 groups of blue hats" task). How are today's comparison problems similar to, but different from, these equal groups problems?

(Let students discuss and explain. T then highlight:)

- In the equal groups problems, we are not comparing two different quantities; there are only equal groups of the same quantity. However, in the comparison problems, we are comparing the bigger and smaller quantities.
- The comparison problem does involve equal groups. We can see the bigger quantity contains multiples of the smaller quantity.

Case 5b: Teacher Y (Conventional)

1. Review the prior knowledge

(1) **Fact family**

T: Friends, we have learned about fact families! Can you use 3, 6, and 18 to write a fact family? (S will likely come up with the following)

$3 \times 6 = 18$

$6 \times 3 = 18$

$18 \div 3 = 6$

$18 \div 6 = 3$

T: I see someone wrote $3 + 6 = 9$. Does it belong to this fact family? Why or why not? What about $6 \div 3 = 2$? (S: Fact family should have the same three numbers. We should only use 3, 6, and 18 here).

T: What about $6 \div 3 = 18$? It does contain the same three numbers, right? (S: But $6 \div 3 = 2$.) Great job!

(2) **Using fact family to solve equal groups problems**

T: Recall that fact family can help us solve word problems. For the following story problem, which number fact will you choose to solve it?

Tom bought 3 hats from a store. Each costed \$6. How much did Tom spend in all?

S: I will choose both $3 \times 6 = 18$ and $6 \times 3 = 18$. They both give me the same answer 18.

T: Do you all agree? (Yes!)

T: Great! Now, what about these two questions?

Tom spent \$18 to buy hats. Each hat costs \$6. How many hats did Tom buy?
Tom spent \$18 to buy 3 hats from a store. How much does each hat cost?

S: I choose $18 \div 6 = 3$ to solve the second problem.

T: Are there any different ways to solve it? Can we use a multiplication equation with an unknown to solve it? (let students think).

T: Who can read the problem loud? (Ask a student to read).
T: This time, can you highlight the key words on your own? (Let students work on it). Which information did you highlight?
S: I highlighted \$18, 3 times, and blue hat.
T: Great! So, how can we find the cost for the blue hat?
S: It is \$6!
T: how do you know?
S: We know the cost of the blue hat from the previous problem!
T: Good thinking! Now, can we write an equation to solve it?
S: $18 \div 3 = 6$
T: Excellent! And how do you know that $18 \div 3$ equals 6?
S: I know $3 \times 6 = 18$, so $18 \div 3 = 6$.
T: You have a good understanding of the relationship between multiplication and division.
T: Now, why did we use division to find the cost of the blue hat? Why didn't we multiply?
S: Blue the blue hat costs less, and division gives us a smaller number.
T: Exactly! Multiplication gives you a big number, while division gives you a smaller number. That's the secret about when to multiply and when to divide. Got it? (Yes!).

4. Compare the equations to complete a fact family

T: Let's take a look at the equations. For the first word problem, what is our equation? ($6 \times 3 = 18$).
T: And for the second word problem, what is our equation? ($18 \div 3 = 6$)
T: Do you see how these two number facts are related?
S: They have the same three numbers! They are part of the fact family.
T: Correct! So, what are the other two missing number facts in this fact family?
S: $3 \times 6 = 18$; $18 \div 6 = 3$
T: Excellent! This is the fact family we reviewed at the beginning of the lesson. So, we can also use it to solve the comparison word problems!

STEP 3: STOP AND REFLECT

Based on your reading and annotating Cases 5a and 5b, please reflect on the following:

1 **Representation uses.** Compare the representation uses of Teachers X and Y. How do their representational sequences differ? How are real-world contexts used differently?
2 **Deep Questioning.** Compare the deep questions asked by Teachers X and Y. How do they differ in terms of frequency and purpose of using deep questions?
3 **Linking to ICAP.** Based on the ICAP framework, which overall level of student engagement does each teacher likely promote? How might each teacher's lesson lead to the respective level?
4 **Insights Gained.** Compare your initial plan with the contrasting cases. What new insights have you gained that you can apply to your lesson planning and future teaching?

4.3.3 *Key to Analysis of Case 5*

4.3.3.1 *An Overview*

The two lessons in Case 5 share similarities. Both Teacher X and Teacher Y aimed to teach the concept of the multiplicative inverse using a comparison model, incorporating Common Core story problems (solved by $3 \times 6 = 18$, $18 \div 3 = 6$, and $18 \div 6 = 3$). Both lessons employed concrete story problem contexts, semi-concrete cubes, and abstract number sentences. However, the foci of the two lessons differed. Case 5a emphasized meaning-making of the number sentences and the inverse relationships through real-world contexts, highlighting quantitative relationships, whereas Case 5b emphasized computation, with real-world contexts serving primarily as applications of the fact family. This was evident in its emphasis on key words and certain tricks rather than on quantitative relationships, which led to mistakes. Although both lessons mentioned the inverse relationship, Case 5a illustrated it seamlessly through the real-world context, while Case 5b primarily highlighted it through computation. Table 4.6 below presents the key points and example deep questions, followed by elaborations.

4.3.3.2 *Representation Uses*

Case 5a began with a review task situated in a real-world context about the cost of blue hats. This set of word problems reviewed the equal groups structure and then led into comparison problems used as a worked example. Semi-concrete chips were employed to represent the prices of red and blue hats. By grouping the chips into boxes, the representation evolved into tape diagrams, providing space for reasoning about quantitative relationships. Based on this reasoning, numerical solutions were generated. The representational sequence (word problems → chips → tape diagrams → number sentences) illustrates the process of concreteness fading. It should be noted that in Case 5a, Teacher X made effective use of the tape diagram to help students understand the key concept of "3 times," the quantitative relationships in each word problem, and the inverse relations indicated by these problems.

In Case 5b, Teacher Y started the lesson with abstract number sentences, which were then used to solve to word problems (e.g., "choose one number fact that applies"). This sequence is the reverse of concreteness fading. Later, when solving real-world problems, semi-concrete cubes were used; however, the purpose of the cubes was limited to finding and checking answers. In contrast with Case 5a, the meaning of "3 times" and the underlying quantitative relationships were not addressed. Rather, key words were emphasized, and certain tricks that involved mathematical mistakes were introduced (e.g., "Multiplication gives you a big number, while division gives you a smaller number").

4.3.3.3 *Deep Questions*

Similar to the earlier cases, in Case 5, deep questioning in Teachers X and Y's lessons also differed across three types of questions: problem-posing, meaning-making, and comparison.

Problem-Posing Questions. During the review task, Teacher X provided students with two opportunities to practice problem posing. First, when the picture—three hats, each priced at \$6—was presented, the teacher asked, "*Who can create a multiplication problem*?" After this multiplication problem was discussed, the teacher followed up with, "*What if we know that Tom spent \$18? Can you change this problem into two division problems?*" Such opportunities required students to go

TABLE 4.6 Key Points and Example Deep Questions

	Case 5a	*Case 5b*
Representation uses	A real-world context (hats) → Chips→ Tape diagrams → Number sentences. This sequence shows *concreteness fading*. The tape diagram models the real-world situation and helps students understand the key concept, quantitative relationships, and inverse relations.	Number sentences → Real-world problems → Cubes. This sequence is different from concreteness fading. Real-world problems were used to apply the learned fact family, and cubes were used to find answers. During this process, key words were emphasized and mistakes were introduced.
Deep questioning	**Problem-Posing Questions** • Who can create a multiplication problem? • What if we know that Tom spent $18? Can you change this problem to two division problems? **Meaning-Making Questions** • Why did we use 3 × 6? • Who can explain the meaning of your equations? • Can you explain why you used division? • Can someone explain to us what "3 times" actually means? • Who can help me quickly see that the cost of the red hat is "3 times" as much as the blue? • Can you tell me how it shows "3 times"? • Who can explain this tape diagram based on the hat problem situation? • Why did we draw 3 sections in the second tape? • What does "3 times" mean? • Why do we use multiplication? • How can we draw the second tape to represent the blue hat? • How do you know the tape for the blue hat should be this much long? • What does this problem tell us about the relationship between the red and blue hats? • How can we show "3 times" on the tape? What does "3 times" mean? • Why do we use division? Can you explain your equation based on the diagram? **Comparison Questions** • Let's compare these word problems (equal groups type). What is similar and what is different? When do we use multiplication? When do we use division? • Let's take a look at these two problems (comparisons type). How are they similar and different? When do we use multiplication and when do we use division? • How are today's comparison problems similar to, but different from, these equal groups problems?	**Problem-Posing Questions** • NA **Meaning-Making Questions** • NA • *Missing opportunities.* Among the two student representations (see Figure 4.18), the first one was incorrect, but it could have been used to stress the meaning. **Comparison Questions** • NA • *Missing opportunity.* Students represented the red cubes in two ways (see Figure 4.18), yet, the first was inaccurate. Teacher Y could have asked students to compare the two cube representations.

(*Continued*)

TABLE 4.6 (Continued)

	Case 5a	*Case 5b*
Linking to ICAP	This lesson likely promoted students' cognitive engagement up to the C or I level. This was evident in Teacher X's representation uses and deep questioning described above.	This likely prompted students' cognitive engagement up to the A-level. This was evident in Teacher Y's representation uses and the absence of deep questioning described above.
Insights Gained	Answers may vary.	

beyond the given material and make new inferences. Consistent exposure to problem-posing opportunities is likely to promote deep learning. In contrast, in Case 5b, no opportunities for problem-posing were provided.

Meaning-Making Questions. As seen in Table 4.6, Teacher X asked a series of questions prompting students to articulate the meaning of "3 times." (e.g., *Can someone explain to us what "3 times" actually means? Who can help me quickly see that the cost of the red hat is "3 times" as much as the blue? Can you tell me how it shows "3 times*"?). As noted in Section 4.3.1, researchers (Riley & Greeno, 1988; Stern, 1993) have pointed out that comparison word problems are relatively abstract because they deal with relationships that cannot be physically manipulated. As such, it is critical to help students understand key concepts like "3 times," which can be linked to the basic meaning of multiplication (three groups of something being compared). In Case 5a, Teacher X consistently stressed the meaning of "3 times" when discussing each type of comparison word problem (finding the multiple, finding the bigger, finding the smaller). Moreover, Teacher X encouraged students to explain meanings through tape diagrams (e.g., *Who can explain this tape diagram based on the hat problem? Why did we draw three sections in the second tape? Why do we use division? Can you explain your equation based on the diagram?*).

In contrast, Teacher Y in Case 5b did not explicitly stress the basic meaning of multiplication or the concept of "3 times." For instance, during the review task, "3 groups of 6" was interchangeably represented as either 3×6 or 6×3 without clarification. Later, when working through the worked example task, the meaning of "3 times" was not addressed. Instead, Teacher Y guided students to rely on a "key words" strategy (e.g., *Can you highlight the key words on your own?*). For instance, the key words "3 times" were linked to multiplication (e.g., *Yes, the key words "3 times" tell us to use 6×3*), which is not always correct and may cause confusion. Moreover, when students suggested an equation (e.g., $18 \div 3 = 6$), instead of asking why they used division, Teacher Y moved on to a discussion of computation (e.g., *And how do you know that $18 \div 3$ equals 6?*).

On a few occasions, Teacher Y asked potentially deep questions but accepted incomplete or wrong responses. For example, when asking "*How should you know 6×3 is 18*?" the teacher accepted the student response "$6 + 6 + 6$," which corresponds to 3×6, not 6×3. Similarly, when asking "*How would you make a cube train for the cost of the red hat?*" and two different representations were offered, both were accepted, even though the first was incorrect. Teacher Y also asked "*Why did we use division to find the cost for the blue hat? Why didn't we multiply?*"—a strong meaning-making question. However, the student's superficial answer ("*Blue hat has less money. Only when you divide, you can get a small number.*") was confirmed and even elaborated by

Teacher Y ("*Excellent, multiplication gives you a big number; division gives you a small number. That's the secret about when to multiply and when to divide.*"). As mentioned earlier, this trick involved mathematical mistakes, because multiplying or dividing by a number less than 1 leads to the opposite conclusion. Finally, during the review, Teacher Y introduced a counterexample of a fact family. Yet, as in earlier cases, the emphasis remained on "the same three numbers," which is a surface-level and potentially misleading observation.

Comparison Questions. In Case 5a, comparison questions were systematically used with clear purpose. This included a comparison of the set of review tasks, where were equal groups problems, leading to a summary of the quantitative relationships. (*Let's compare these word problems. What is similar and what is different? When do we use multiplication? When do we use division?).* In the new lesson, Teacher X guided students to compare two newly learned comparison problems (*Let's take a look at these two problems. How are they similar and different? When do we use multiplication and when do we use division?).* After the worked example is completed, the teacher further asked the class to compare the new teaching with their prior knowledge. (*How are today's comparison problems similar to, but different from, these equal groups problems?*). These comparison questions consistently prompted students to make connections between various knowledge points, which likely deepened their understanding of the new concepts.

Case 5b did not include comparison questions. Even though there were potential opportunities to highlight inverse relations, Teacher Y did not prompt students to compare or articulate them. For instance, regarding the two student representations mentioned earlier, Teacher Y could have asked students to compare them, which might have deepened their understanding of the concept of "3 times." Additionally, when Teacher Y asked, "How do you know $18 \div 6$ is 3?" students responded, "I know $3 \times 6 = 18$, so $18 \div 6 = 3$." Teacher Y only praised the response with a quick remark: "Great! Multiplication can help us solve division." In fact, Teacher Y could have followed up by asking how multiplication and division are related. Moreover, Teacher Y taught the same three comparison word problems about red and blue hats, but could have asked similar comparison questions at various levels to prompt student reasoning and inference-making, as Teacher X did.

4.3.3.4 Linking to ICAP

From the perspective of ICAP, we can conclude that, overall, Teacher X's lesson in Case 5a prompted students' cognitive engagement up to at least the C level, while Teacher Y's lesson in Case 5b remained at the A level. This difference is evidenced by the teachers' contrasting uses of representations and questioning when addressing the same concept.

In Case 5a, representations (tape diagrams) served as a platform for deep questioning that elicited students' understanding of quantitative relationships in comparison word problems and the multiplicative inverses embedded within them. Answering the deep questions raised by Teacher X required students to extend their reasoning beyond the given materials and make new inferences throughout the lesson—an indicator of C-level cognitive engagement.

In contrast, in Case 5b, Teacher Y asked students to use cube representations primarily for answer-seeking. Word problem contexts were treated merely as a pretext for computation – using a fact family to help solve word problems without discussing the underlying quantitative relationships. In other words, Teacher Y did not ask students to pose problems, make sense of numerical solutions based on cube representations, or compare different solutions and problems to uncover structural connections. As such, opportunities for deep learning were missed. In this sense, we conclude that Teacher Y's lesson remained at the A level.

4.3.3.5 Insights Gained

My pilot work with Case 5 in the methods class shows that PSTs had different takeaways, but they commonly recognized the strengths of the TEPS approach, especially in terms of using representations and deep questions. One PST pointed out that Teacher X's use of representations helped unpack the concepts, which even enhanced her own understanding. This PTS wrote:

> By comparing both teachers, Teacher X is the type of teacher I want to be like. I thought their use of varied representations was very effective in giving children the chance to see the concepts. Even while I was reading Teacher X's lesson, I could really see the connection between all the concepts and how the numbers can be broken up

PSTs were also impressed by Teacher X's deep questioning. At the same time, a couple of PSTs expressed concerns about Teacher Y's emphasis on key words. As one PST noted, Teacher X asked more questions that helped students understand meaning—for example, discussing the relationships between the different color hats or asking students to show their thinking—while Teacher Y focused mainly on key words.

Interestingly, in noticing the overall conceptual emphasis in Teacher X's lesson and the procedural focus in Teacher Y's lesson, a few PSTs called for a balanced view between the two approaches. As mentioned in earlier chapters, while I agree that a balanced view is important when teaching a "full" lesson, I emphasize conceptual underpinnings when teaching the worked-example portion of a lesson. In this regard, the purposes of the contrasting cases could have been made more explicit to PSTs. Nevertheless, PSTs in my course generally grasped the intended instructional insights, as illustrated by the following quote:

> Encouraging students to explore and discuss their thinking, as seen in Teacher X's lesson, can lead to deeper understanding and stronger problem-solving skills, compared to the structured, step-by-step guidance in Teacher Y's approach. Teacher X made clear connections between different types of comparison problems, reinforcing the relationship between multiplication and division, while Teacher Y connected ideas through fact families but with less emphasis on conceptual depth. Moving forward, lesson planning should include hands-on activities, a mix of exploration and structured practice, opportunities for students to explain their reasoning, and a variety of problem types to help students develop a strong, flexible understanding of mathematics

4.4 Summary

In this chapter, I presented three pairs of Cases (3a and 3b, 4a and 4b, 5a and 5b) to illustrate how one might approach a worked example of multiplicative inverse, incorporating three typical multiplication models: equal groups, array, and comparison. Each pair of cases demonstrated either the TEPS approach or a conventional approach. The worked-example tasks were adapted from the Common Core State Standards, which increases the potential transferability of these cases, because most elementary textbooks in the United States are aligned with the Common Core.

Across the three cases, Teacher X used the TEPS approach, situating the worked examples in real-world contexts (bags of plums, arrays of apples, and the costs of blue and red hats). By presenting such pictorial contexts, students may be immediately engaged and motivated to pose questions. Incorporating students' own contributions into new teaching can also give them a stronger sense

STEP 1: THINK AND PREPARE

Imagine you are a first-grade teacher. How would you teach the commutative property of addition (e.g., $3 + 5 = 5 + 3$) based on a part-whole model? Draft a brief plan in a T–S dialogue format. What representations would you use, and how would you sequence them? What questions, including deep questions, would you ask, and what student explanations would you anticipate?

Your Initial Draft Plan:

5.1.2 *Contrasting Cases: 6a and 6b*

In Case 6, two subcases illustrate how first-grade teachers approach a worked example of the commutative property of addition (not a full lesson). Case 6a features Teacher X using the TEPS approach, while Case 6b showcases Teacher Y employing a conventional method. Notably, students in both classes possess the same prior knowledge, and the two lessons share the same instructional objectives.

Prior Knowledge. Students have already learned how to solve part-part-whole word problems.

Teaching Objectives. Students will be able to:

1 Make sense of the commutative property of addition
2 Apply commutative property of addition for computation

STEP 2: READ AND ANNOTATE

Below are Cases 6a and 6b. Read and annotate each case, focusing on the use of representations and questioning. Specifically, for each Teacher (X and Y), identify:

a the types of representations used
b the deep questions asked

Case 6a: Teacher X (TEPS)

1. Teach two ways to solve a word problem

T: Friends, do you like swimming? (S: Yes!). Look, this is a swimming pool (Show the picture in Figure 5.1). What do you notice about this swimming pool?

S: (Let a few students share.) Three children are on the bank, and five are in the pool.

T: Can you pose a math question?

(Students may give the answer or state the equation. The teacher should guide them to pose a question.)

T: Who can state the question completely?

S: There are 3 children on the bank of the pool and 5 in the pool. How many children are there in total?

FIGURE 5.1 The swimming pool context. This image was taken from Sun & Wang (2021, v1, p.58). Permission of citation was granted.

T: Let's use counters to represent the children. Can you use two different colors to show the children on the bank and the ones in the pool? (Let students come to the board to represent it. See Figure 5.2.)

T: What does these red counters represent? (Children on the bank). What do the yellow counters represent? (Children in the pool).

T: How can we find how many children there are in total?

S: $3+5=8$

T: What does this equation mean? (There are 3 children on the bank and 5 in the pool, making a total of 8 children.)

T: In this problem, 3 is one part, and 5 is the other part. We add them together to find the total.

T: We just looked at the children on bank first. Is there another way we can look at this picture? (we can look at the children in the pool first).

T: So, if we use counters, we can start with 5 yellow counters and then 3 red counters (show the counters).

FIGURE 5.2 A counter representation for the swimming problem.

T: How can we write an equation to find the total number of children?
S: 5+3=8
T: (Write "5 + 3 = 8" underneath "3 + 5 = 8"). Can you explain to your deskmate what 5 + 3 = 8 means? (Ask one student to share with the class). In this problem, we are still adding two parts to find the total.

2. Compare the two solutions

T: For this same story problem, we just solved it in two ways. We can start with the children? (Ss: on the bank) or start with the children? (Ss: in the pool). That gives us two equations: 3 + 5 = 8 and 5 + 3 = 8.
T: Compare these two addition equations. What do you notice? Discuss with your desk mates.
S: (Ask a few students to share) 5 and 3 are switched, but the answer remains the same. The total is still 8.
T: Why do both equations give us the same answer?
S: We just switched the two parts around, but the total hasn't changed.
T: Excellent! When we add 3 and 5, we can switch the two numbers, and the answer stays the same. What sign should we put here? (Points to the equation: 3 + 5 ○ 5 + 3)
S: The equal sign!
T: Exactly! So, 3 + 5 = 5 + 3

3. Other ways to look at this picture

T: We just separated these children into those on the bank and those in the pool. We solved this addition problem in two ways. Are there any other ways that we can separate these children?
S: We can separate them into boys and girls
T: Can you state your story problem completely?
S: There are 5 boys and 3 girls swimming. How many children are there in total?
T: Can we solve this problem in two ways?
S: 5+3=8 and 3+5=8.
T: We got the same equations as we discussed before. What about their meanings? Are they the same? (No, they have different meanings).
T: Do we have any other ways to separate these children into different groups?
S: We can separate them according to their lifebuoys. Seven children's lifebuoy are circles; but one child's lifebuoy has a duck.
T: Your way is very special. Based on this, what two addition equations can we write? (7 + 1 = 8, 1 + 7 = 8).
T: What's the relationship between 7 + 1 and 1 + 7? (7 + 1 = 1 + 7)
T: Are there any other ways to look at this picture?
S: Six lifebuoys have strips, while two lifebuoys do not.
T: This is also special. Who can come up with two equations based on this? Can you tell your desk mates?
S: (Ask one student to share.) 6 + 2 = 8 and 2 + 6 = 8
T: What sign can we use to connect 6 + 2 and 2 + 6?
S: The equal sign. (T: write: 6 + 2 = 2 + 6).

4. Identify the pattern and reveal the commutative property

T: We can observe one picture from different angles and come up with different solutions. Let's look at these pairs of equations. What do you notice that is common across these equations?

3+5=5+3

7+1=1+7

6+2=2+6

S: (Let different students express their opinions. Possible responses:

- They are all addition equations.
- We added two parts on the left side, and we switched them on the right side.
- We switched the numbers, but they are still equal.)

T: Excellent. When we switch the two parts to add, the total does not change. In other words, when we add, if we switch the two numbers, the answer will not change. We call this big idea the commutative property of addition. Read together two times, "the commutative property of addition." (Writes the lesson title on the board).

S: (Read). "The commutative property of addition."

T: If I use Δ to represent one number and ♦ to represent another number, Δ + ♦ would be the same as what? Discuss in your desk mate. (♦ + Δ)

T: Excellent! You all did a great job.

Case 6b: Teacher Y (Conventional)

1. Introduce the definition of the commutative property

T: Today, you'll be learning about the commutative property. Everyone says "commutative."

Ss: Commutative.

T: Right. The commutative property states that the order in which the numbers are added does not change the sum. You can think of fact families, or turnaround facts, when discussing the commutative property. For example, $4 + 6 = 10$. The commutative property tells me that if I know $4 + 6 = 10$, I can switch the order of these two addends, and I know that $6 + 4 = 10$. So, the order in which the numbers are added does not change the answer. The sum remains the same.

2. Illustrate with an example using counters

T: Why is it important to understand these properties of addition? Knowing these properties will make our addition easier. Let's look at this example from our textbook. It says to place 1 to 5 red counters in the left box and 6 to 9 yellow counters in the right box. Then, write the addition sentence represented by the parts.

T: So I'm going to put two red counters here (drags 2 red counters into one box on the smartboard) and six yellow counters here (drags 6 yellow counters into the other box, counting aloud). So if I have 2 red and 6 yellow (see Figure 5.3), what addition sentence can I write to show how many counters I have in all?

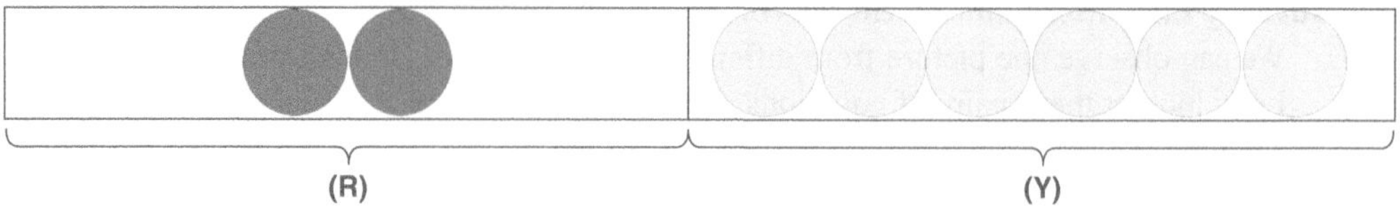

FIGURE 5.3 Two red and six yellow counters

S: 2+6

T: Good. 2 + 6. And what is 2 + 6?

S: 8

T: 8. All right. So, 2 + 6 = 8.

T: Now look at this. Place as many red counters in the left box as there were yellow counters on the previous screen. So, before we used six yellow, so now we're going to do six red (drags red counters counting aloud). Place as many yellow counters in the right box as there were red counters on the previous screen. Since I had two red ones before, I'm going to use two yellow now (drags counters, counting aloud again) (see Figure 5.4).

T: So now, what would my addition sentence be if I'm adding this many red and this many yellow?

S: 6+2

T: Good. Now we have 6 + 2. And what is 6 + 2?

S: 8

T: 8. So what did you notice about the one we just did, where I had two yellow counters and six red, and the one I have on the board now? What do you notice about those addition sentences?

S: The answer is the same.

T: The answer is the same; the sum is the same. It didn't matter that we switched the order of the addends. 2 + 6 is the same as 6 + 2. And that is an example of the commutative property. No matter the order of the addends, the sum doesn't change.

3. Reinforce with another example using blocks

T: Let's look at a story problem. One day, a boy has 5 blocks of the same color, and then he adds 6 more blocks of a different color. How many blocks does he have? This is what I want you to show me. Go! 5 blocks plus 6 blocks. (T walks around to check)

T: (show the class) Put the blocks together—5 blocks and 6 blocks. Look at this (see Figure 5.5). When you have them like this… Stand up at your table.

T: Ready? We're going to turn them around. Ready? So, 6 blocks and 5 blocks (see Figure 5.6).

T: What changed?

S: The numbers, but it's still the same.

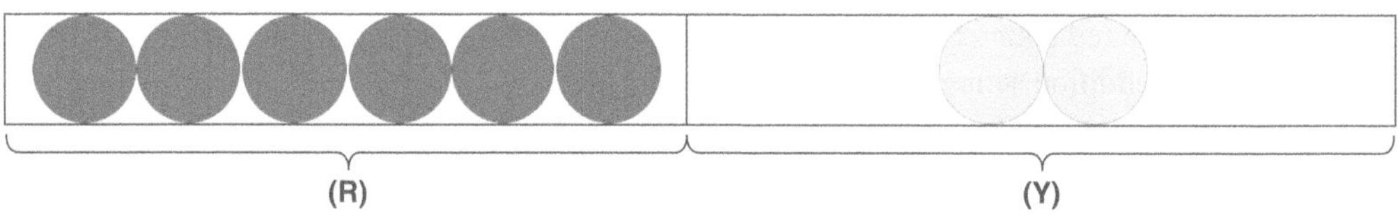

FIGURE 5.4 Six red and two yellow counters

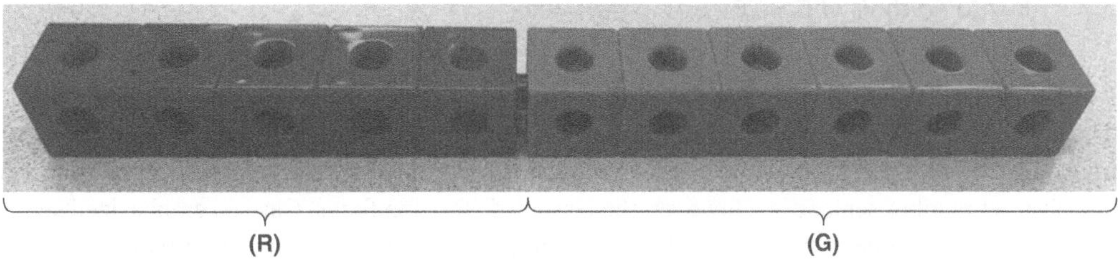

FIGURE 5.5 Block representation for 5 + 6

FIGURE 5.6 Block representation for 6 + 5

T: That's right. How many are here? (Refer back to the original blocks).
S: 11.
T: So, 5 + 6 = 11.
T: Now watch. (Teacher turns the blocks around). How many are here?
S: 11.
T: So, 6 + 5 = 11. I see that the order of my addends switched, but my sum stayed the same. So which property does this show?
S: Commutative
T: Good, the commutative property. So, the order in which you put the addends doesn't change the sum. That's what we call turnaround facts. Great job, everyone!

STEP 3: STOP AND REFLECT

Based on your reading and annotating Cases 6a and 6b, please reflect on the following:

1 **Representation uses.** Compare the representation uses of Teachers X and Y. How do their representational sequences differ? How are real-world contexts used differently?
2 **Deep Questioning.** Compare the deep questions asked by Teachers X and Y. How do they differ in terms of frequency and purpose of using deep questions?
3 **Linking to ICAP.** Based on the ICAP framework, which overall level of student engagement does each teacher likely promote? How might each teacher's lesson lead to the respective level?
4 **Insights Gained.** Compare your initial plan with the contrasting cases. What new insights have you gained that you can apply to your lesson planning and future teaching?

5.1.3 Key to Analysis of Case 6

5.1.3.1 An Overview

The two lessons in Case 6 share several similarities. Both aimed to teach the commutative property of addition using a part-whole model and employed story problems, semi-concrete cubes/counters, and abstract number sentences. However, the lessons differed in focus. Case 6a emphasized sense-making of the number sentences and the commutative property through real-world contexts, whereas Case 6b emphasized computation, with the story problems serving primarily as an application. Table 5.1 below highlights key points and provides example deep questions.

5.1.3.2 Representation Uses

In Case 6a, Teacher X began the lesson with a real-world context—a swimming pool. Students were encouraged to observe this picture and pose story problems. The context was then represented with semi-concrete counters and ultimately solved using abstract number sentences. This sequence demonstrates concreteness fading. During teaching, each equation was explicitly linked back to the story context to support meaning-making. Additionally, the same real-world context was interpreted in multiple ways (e.g., 3 on the bank & 5 in the pool; 3 girls and 5 boys; 7 children's lifebuoys as circles & 1 child's lifebuoy with a duck; 6 lifebuoys with stripes & 2 without). These different perspectives generated a set of equations (e.g., $3 + 5 = 5 + 3$, $1 + 7 = 7 + 1$, $2 + 6 = 6 + 2$), providing opportunities for students to explore the pattern and discover the commutative property. This representation highlights its purpose in facilitating understanding of quantitative relationships and overarching mathematical ideas beyond mere numerical facts.

In Case 6b, Teacher Y began the lesson with an abstract definition of the commutative property, illustrated with number facts and an example using counters. This sequence contrasts with concreteness fading. Because the commutative property was directly presented, students' learning was relatively passive, as they were not prompted to generate new information. Although counter illustrations were included, the meaning of the numbers changed arbitrarily (e.g., "2 yellow counters" became "2 red counters"), limiting their effectiveness in showing how $2 + 6 = 6 + 2$. Additionally, a story problem was presented at the end; however, it involved only blocks, and the resulting number sentences were not sufficiently discussed. Students primarily engaged in counting and computation, so while physical engagement was present, opportunities for inference-making and deeper understanding were largely absent.

5.1.3.3 Deep Questions

The use of deep questioning in Teachers X and Y's lessons differed across three types: problem-posing, meaning-making, and comparison.

Problem-Posing Question. Teacher X in Case 6a asked several problem-posing questions. When the swimming pool picture was presented, the teacher invited students to observe it and pose a math question (e.g., "*Can you pose a math question? Who can state this question completely*?"). Later, the teacher invited students to view the picture differently (e.g., "*Is there another way we can look at this picture?*" "*Are there other ways we can separate these children*?"). These questions encouraged students to examine the same picture from multiple perspectives. As a result, students exercised their creative thinking skills and generated different story problems, providing a foundation for reasoning and sense-making. These discussions enabled students to identify the

TABLE 5.1 Key Points and Example Deep Questions

	Case 6a	*Case 6b*
Representation uses	Real-world context → Counters→ Number sentences. This sequence illustrates concreteness fading. The real-world context (swimming pool) was explored in depth to support meaning-making and to help students identify the commutative property.	Definition → Number facts→ Counter→ Story problem about blocks. This sequence contrasts with concreteness fading. The counter illustration offered little support for sense-making and was used primarily for computation. The so-called story problem (about blocks) reinforced what was already taught.
Deep questioning	**Problem-Posing Questions** • Can you pose a math question? Who can state this question completely? • Is there another way we can look at this picture? • Are there any other ways that we can separate these children? Can you say your story problem completely?	**Problem-Posing Questions** • NA
	Meaning-Making Questions • What does this equation mean? • We got the same equations as what we just discussed. What about the meanings? Are they same?	**Meaning-Making Questions** • NA
	Comparison Questions • Compare these two addition equations, what do you find? • Why do both equations give us the same answer? • Let's look at these pairs of equations. What are common across these equations?	**Comparison Questions** (*but the anticipated answers remained at surface level*) • What do you notice about those addition sentences? (The answer is the same.) • What changed?
	Other ("what if" question) • If I use Δ to represent one number and ♦ to represent another number, Δ + ♦ would be the same as what? Discuss in your desk mate.	**Other (question on "purpose")** • Why is it important to understand these properties of addition? Knowing these properties will make our addition easier for us.
Linking to ICAP	Promotes students' understanding at least to the C level and sometimes to the I level. This is evident in Teacher X's use of representations and deep questions as described above.	Understanding is mainly limited to the P or A-level. This is evident in Teacher Y's use of representations and the absence of deep questions as described above.
Insights Gained	Answers may vary.	

commutative property, the targeted concept of the lesson. One can imagine that students consistently exposed to such teaching would benefit from inference-making opportunities beyond the given materials (Cai, 1998; Cai & Rott, 2024; Cai et al., 2023; Silver & Cai, 1996).

In contrast, Teacher Y in Case 6b did not prompt students to pose any questions. This may be because the representations (counters, and a story problem about blocks) in Case 6b were used primarily to support computation and demonstration rather than to stimulate problem solving exploration or reasoning.

Meaning-Making Questions. Teacher X in Case 6a also asked students to explain the meaning of the equations (e.g., "*What does this equation mean?*"). Interestingly, because students interpreted the same picture in different ways (e.g., 3 on the bank & 5 in the pool; 3 girls & 5 boys), the same equation, $3 + 5 = 5 + 3$, carried different meanings. Teacher X prompted students further: "*We got the same equations as we just discussed. What about the meanings? Are they the same?*" Such questions encouraged students to think beyond the equations and facilitated connections between the equations and the story problems.

In contrast, these types of deep questions were completely absent from Case 6b. Teacher Y's questions focused primarily on computational answers (e.g., "*How many are here*?") or on comparing these answers (see below), rather than on developing conceptual understanding.

Comparison Questions. Both teachers asked comparison questions. Teacher X in Case 6a not only asked students to compare (e.g., "*Compare these two addition equations. What do you find?*") but also encouraged them to consider the reasons behind their observations (e.g., "*Why do both equations give us the same answer?*") and to explore patterns based on their observations (e.g., "*Let's look at these pairs of equations. What is common across these equations?*"). To understand the reasons, students were expected to refer back to the swimming pool context and recognize that only the two parts were switched while the whole remained unchanged—a sophisticated explanation for first graders. Later, by examining the generated equations, students were expected to verbalize the commonality, leading to the revealing of the commutative property. This contrasts with Case 6b, where the commutative property was directly introduced to students at the beginning of the lesson.

Teacher Y in Case 6b also asked two comparison questions. However, in both cases, responses remained at a surface level, focusing primarily on the answers. For example, after presenting the counter representations, Teacher Y asked students what they noticed about the two representations. Although this was a deep question that could promote comparisons, the teacher accepted students' brief responses about the same answer, missing opportunities to guide further discussion on patterns. Similarly, at the end of the lesson, the teacher switched the blocks and asked, "What changed?" A student responded, "The numbers, but it's still the same." This observation could have been further prompted for articulation. However, Teacher Y directly stated the commutative property to the students, which limited the potential of the comparison questions and highlighted the need for follow-up questions like those used by Teacher X.

Other Questions. In both lessons, there were other interesting questions worth highlighting. In Case 6a, Teacher X asked a "what if" question: "*If I use Δ to represent one number and ♦ to represent another number, Δ + ♦ would be the same as what? Discuss in your desk mate.*" This age-appropriate, deep question likely activated students' interest and fostered generalization skills through peer discussion. In Case 6b, Teacher Y asked a question about the purpose of learning this property: "*Why is it important to understand these properties of addition?*" While this is a good question, the teacher immediately provided the answer, stating "*Knowing these properties will make our addition easier for us*," without giving students an opportunity to think independently.

5.1.3.4 Linking to ICAP

From the perspective of ICAP (Chi & Wylie, 2014; Chi et al., 2018), Teacher X's lesson guided students to identify the commutative property through a problem-solving process, thereby promoting deeper levels of learning. As analyzed above, the real-world swimming pool context was used effectively, supported by problem-posing, meaning-making, and comparison questions. Together, these instructional approaches likely encouraged students to move beyond computing basic facts to reasoning about underlying relationships and making various inferences. Accordingly, students' cognitive engagement level in Case 6a at least reached the C level. Since Teacher X also facilitated peer discussion around challenging questions, it is reasonable to conclude that students' learning may even have reached the I level.

In contrast, Teacher Y in Case 6b began by introducing the commutative property and then illustrated it with numerical examples and counter/block representations. This reduced the need for reasoning and for actively identifying the property. Students in Teacher Y's lesson were mainly engaged in counting and computing. When asked to compare the generated equations, they were not pushed beyond the visible information (e.g., "*The answers are the same*"). As a result, new inference-making was rare in Case 6b. Based on this evidence, students' engagement in this lesson was likely limited to the P or A level.

5.1.3.5 Insights Gained

After reading and comparing Case 6, PSTs in my math methods course commonly agreed that these contrasting cases offered valuable insights for their future teaching practice. They especially favored the concreteness fading approach, particularly the use of real-world contexts to support students' learning. One PST commented:

> Teacher X's approach shows the impact of starting with concrete representations to introduce abstract mathematical concepts. By using relatable visuals like a pool scene before transitioning to counters, Teacher X helps students make sense of math in ways that feel tangible and connected to their everyday experiences. This concrete-to-abstract progression appears more effective, especially for early learners, than starting with abstract concepts, as Teacher Y did. It makes me consider how I might use familiar, real-life examples as entry points to help students grasp more complex ideas later

PSTs also commonly valued Teacher X's use of deep questions in developing students' critical thinking. They noted that Teacher Y's questions only requested simple answers, which did not prompt deep learning. A typical response was:

> The differences in questioning techniques also reveal a lot about how to encourage critical thinking. Teacher X's open-ended questions, like "What does this equation mean?" and "Why do both questions give the same answer?" prompt students to explore different ways of understanding and to think deeply about the material. In contrast, Teacher Y's questions, which often had simple answers, kept students more passive. This comparison highlights how essential it is to go beyond yes-or-no questions and instead ask questions that require students to explain their thinking and connect with the material on a deeper level. I realize that to foster an engaging, interactive classroom, I need to incorporate more "why" and "how" questions that drive students to think critically

Finally, some PSTs highlighted that both the *"real-world scenarios"* and *"deep questions"* were key takeaways, as indicated below:

> One key insight I can apply to my lesson planning is the importance of using real-world scenarios to help students connect with the content. Watching how Teacher X engaged her class by having them split children in a pool into different groups showed me how effective it can be to let students take the lead in creating their own learning. This taps into the Interactive level of the ICAP framework, where students are actively generating ideas and making sense of the math themselves. In my future teaching, I want to include more open-ended questions and scenarios that encourage multiple responses, so students can explore different strategies and feel more ownership over their learning. I also realized that while using manipulatives like chips and blocks is helpful, it's not always enough to push students beyond the active level unless it's paired with deeper thinking tasks

Overall, Case 6 generated strong interest among PSTs, who demonstrated meaningful learning from these contrasting lessons.

5.2 Teaching Associative Property of Addition

5.2.1 Introduction

The Common Core State Standards expect students in Grades K–2 to apply the associative property (AP) of addition, along with other basic properties, in computation. For example, the standards state: "*Apply properties of operations as strategies to add and subtract. Examples: (...) To add 2 + 6 + 4, the second two numbers can be added to make a ten, so 2 + 6 + 4 = 2 + 10 = 12. (Associative property of addition.)*" The "make-a-ten-to-add" strategy is a powerful application of the AP. For instance, a fact like $8 + 4$ can be decomposed and recomposed using the AP to form a ten. Beyond basic facts, this strategy generalizes to broader contexts, such as make-tens or make-hundreds when adding larger numbers. One example is $98 + 14 = 98 + (2 + 12) = (98 + 2) + 12 = 100 + 12$. Because of this wide applicability, it is critical that when students first learn the *make-a-ten-to-add* strategy, they not only practice it procedurally but also develop an understanding of how and why the strategy works. In Case 7, we will examine teaching approaches for introducing the make-a-ten-to-add strategy in Grade 1–2 classrooms.

STEP 1: THINK AND PREPARE

Imagine you are a second-grade teacher. How would you teach the associative property of addition through a lesson on make-a-ten-to-add? For example: $9 + 4 = 9 + (1 + 3) = (9 + 1) + 3 = 10 + 3 = 13$. Draft a brief plan in a T–S dialogue format. What representations would you use, and how would you sequence them? What questions, including deep questions, would you ask, and what student explanations would you anticipate?

Your Initial Draft Plan:

5.2.2 Contrasting Cases: 7a and 7b

In Case 7, two subcases illustrate how second-grade teachers approach a worked example of the associative property of addition (not a full lesson). Case 7a features Teacher X using the TEPS approach, while Case 7b showcases Teacher Y employing a conventional method. Both lessons involve teaching the make-a-ten-to-add strategy, during which the associative property of addition is implicitly applied. Students in both classes have the same prior knowledge, and the two lessons share the same teaching objectives.

Prior Knowledge. Students have already learned:

1 How to decompose a number up to 10
2 Addition and subtraction facts up to 10
3 10 plus a number

Teaching Objectives. Students will be able to:

1 Make sense of the make-a-ten-to-add strategy
2 Differentiate between more and less effective strategies for making a ten to add

STEP 2: READ AND ANNOTATE

Below are Cases 7a and 7b. Read and annotate each case, focusing on the use of representations and questioning. Specifically, for each teacher (X and Y), identify:

a the types of representations used
b the deep questions asked

Case 7a: Teacher X (TEPS)

1. Review 10 + Something

T: (Show a cartoon Monkey). Friends! Look who I brought with me today! (Monkey!) Yes, Mr. Monkey has brought some questions for you! (Show the cards one by one: 10 + 4 =, 10 + 7 =, 10 + 5 =, 10 + 2 =, 10 + 6 =, 10 + 3 =, 10 + 8 =)

S: (Students respond quickly).

T: Wow! How did you figure them out so quickly?

T: 10 plus something gives you a ten-something or a teen number! It's easy!

2. Introduce the story problem context and solve it

T: You have learned so well. Mr. Monkey has brought some apples for you as rewards! Take a look — how many apples are inside the box, and how many are outside the box? (see Figure 5.7).

S: There are 9 inside the box and 4 outside the box.

T: Can you pose a question so we can solve this problem using addition?

FIGURE 5.7 Monkey and apples. This image was taken from Sun & Wang (2021, v1, p.88). Permission of citation was granted.

S: There are 9 apples inside the box and 4 outside the box. How many apples are there in total?
T: How can we solve it? Can you write a number sentence?
S: (9 + 4 = 13)
T: Why did you use addition?
S: Because we want to put the two parts together.
T: Excellent! Do you all agree the total is 13? (Yes).

3. Explore the calculation method using the ten-frame

T: How did you calculate 9 + 4? Share your strategies with your desk mates. You can use your ten-frame and counters if needed. (T walks around to listen to group discussions).
T: (Put the ten-frame on the board). Okay. Put aside your manipulatives. I use each counter to represent each apple. Can someone come to the board and explain how you figured out the total number of apples?
S: (Move the counters while explaining) There are 9 apples in the box and 4 outside the box. I moved one apple from outside into the box. Now, there are 10 apples inside the box and 3 outside. 10 + 3 = 13. So, 9 + 4 = 13.
T: What a special way! Let me ask: Why did you move 1 apple from those four into the box?
S: This way, there will be 10 apples in the box. 10 plus something is easy! 10 + 3 = 13.
T: In your way, how did you separate your 4 apples? Which two parts did you come up with?
S: 1 apple and 3 apples.
T: So, what did you calculate first? (S: 9 + 1 = 10).
T: And then? (S: 10 + 3 = 13).
T: (Ask the class) Who thought of the same way? Raise your hands! Great.
T: I want to ask one more student to think aloud. (Let the second student explain).

T: So, both students thought of making the 9 into what? (S: 10!).

T: Excellent! I have a question. You can also separate the 4 apples into 2 and 2. Why didn't you move 2 into the box?

S: It would be too full!

T: What if we had enough space? Would you move 2 in then?

S: No, we want to make a 10 to add because 10 plus something is easy. If you move 2 in, you'll make 11, not 10.

4. Record the manipulative process numerically

T: Great. You just made a 10 to add. Now, let's record your method on the board. How did you break apart the 4 (S: 1 and 3)

T: Yes, 4 is broken into 1 and 3 (T records numerically). First, we calculated 9 and 1, which equals? (10). And then we calculated? (10 + 3 = 13). (T records this process on the board as the following, see Figure 5.8)

T: Who can explain this process completely?

(Asks a few students to explain, then has them discuss with their desk mates: "We break 4 into 1 and 3. First, we calculate 9 + 1 = 10, then we calculate 10 + 3 = 13.")

T: Let's say it together (Pointing to the numerical records on the board). "4 is broken into 1 and 3; 9 + 1 = 10, 10 + 3 = 13."

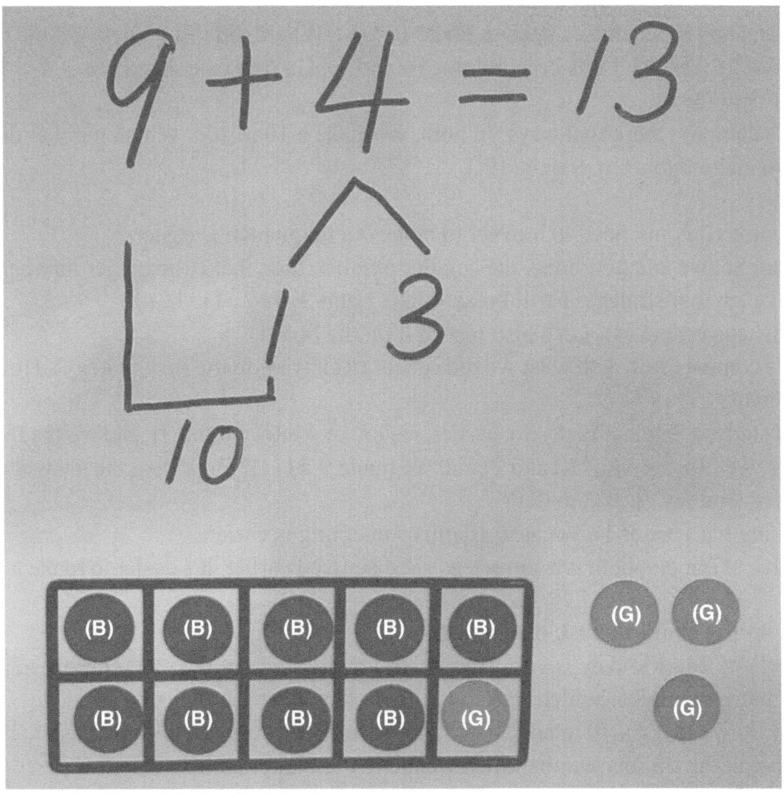

FIGURE 5.8 Numerical representation for 9 + 4 on the board

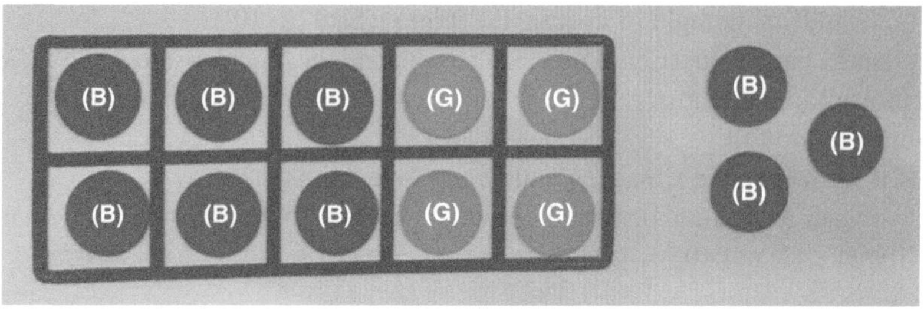

FIGURE 5.9 An alternative way to solve 9 + 4 by turning 4 into 10

5. Explore other ways of calculation

T: Do we have other ways to compute this?

S: I will take out the 3 apples, leaving 6 inside the box. Then, I will move the 4 apples from outside into the box, making 10 apples. Adding the 3 apples outside the box, we have a total of 13 (see Figure 5.9).

T: Do you all understand this strategy? (Yes!) Who can explain what this student said? (Ask another student to explain).

T: Okay. This student also thought of? (Making a 10!). In this strategy, he did not make 9 into 10. Which number did he make a ten? (4!).

T: Great. How much does 4 need to make 10? (6). Where did this student get the 6? (He got 6 from 9). Yes. And 9 is broken into? (6 and 3). He first calculated? (6 + 4 = 10). Then, 10 + 3 = 13.

T: Let's compare these two ways. In both, we make a 10 to add. Which number do you prefer to make into a ten, 9 or 4? (9!)

T: Why?

S: Because you only need to move 1 to make 9 a ten, which is easier.

T: Great. So, we can first break the smaller number, then make the bigger number a 10.

S: I have another strategy. I will break 4 into 2 and 2; 9+2=11, 11+13.

T: Is this correct? (Yes). Let's also record it on the board.

T: Let's compare this with what we did earlier (Referring to the first strategy). How are they different?

T: (Let students explain first.) So, earlier, we broke 4 into? (1 and 3), and we made 9 a? (10); Here, we broke 4 into? (2 and 2) and we made 9 a (11)? Both give the answer of 13.

T: Which strategy do you prefer?

S: Making 9 a 10, not 11, because 10 plus something is easier.

T: I agree. Although both are correct, as we discussed earlier, it's easier to make a 10 to add.

6. Wrap up the example task discussion

T: Excellent. Mr. Monkey is very happy that you figured out the total number of apples using many strategies. Which strategy did you like most? (Make a 10 to add).

T: How do we make a 10 to add? Take a moment to discuss it with your desk mates. (First, we break the smaller number, then we make the bigger number into a 10).

Case 7b: Teacher Y (Conventional)

1. Introduce lesson objective and the example task

T: The objective of today's math lesson is that students will be able to use the strategy of making a ten to add. Let's look at this example problem. (Project the textbook page on the board. See Figure 5.10).

T: Who can read it to us?

S: How can thinking about 10 help you find 9 + 3? Use the ten-frames and counters to show how.

T: What does this problem ask us to do?

S: Thinking about 10 to find 9 + 3; to use counters to show it.

T: Can you give me two addends that make 10?

S: 4 + 6, 5 + 5, 10 + 0, 2 + 8, 7 + 3, and 9 + 1

T: I am going to give out the counters and ten-frames. You can work in small groups or by yourself.

2. Discuss how to make a 10

(Ss work in the seat with the manipulatives while T walks around)

T: Can someone come up and show us how you made a ten to solve 9 + 3? I have two ten frames up here. As you work, could you explain what you are doing?

S: First, I'm putting in 9.

T: (To the class). He is telling me he is putting up how many? (S: 9.)

T: And why are you putting up 9?

S: To make the first part of the equation, 9.

T: Great! So I am going to write 9.

T: Could someone else come to show us the rest? (One S puts up 3 green counters).

Solve and Share:

How can thinking about 10 help you find 9+3?
Use the ten-frames and counters to show how.

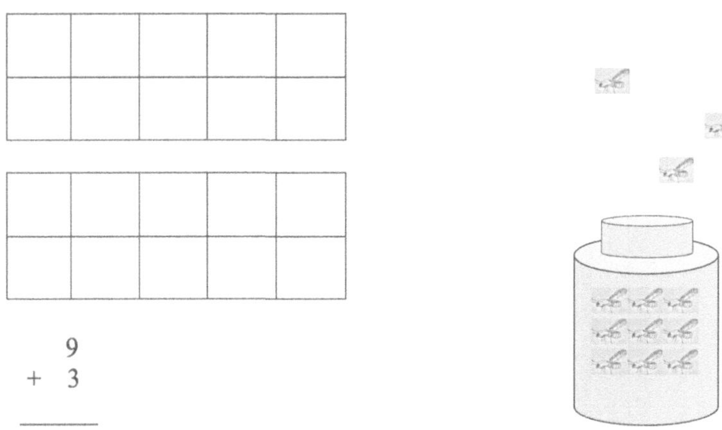

FIGURE 5.10 A modified textbook page for teaching 9 + 3. The firefly is drawn by Owen Li

T: Tell us what you did.
S: I put 3 up.
T: Why did you put 3 up?
S: To put the second part of the equation.
T: (Completing the equation) So 9 + 3, she put up 3 to show the second part of the equation. Then what happens?
S: You move one of these (points to a green counter) into the first frame.
T: Why did you move this over there?
S: To make the ten.
T: To make a ten! That's what we're doing, right?
S: Then we have 10 and 12. 10, 11, 12.
T: So what was the sum of 9 + 3? (S: 12.)
T: 12. And now we have a new equation. Could you tell us? (S: 10 + 2.)
T: 10 + 2. And what is the sum? (S: 12.)
T: 12. So, 10 + 2 = 12 (writes the equation. See Figure 5.11). Give them a round of applause!

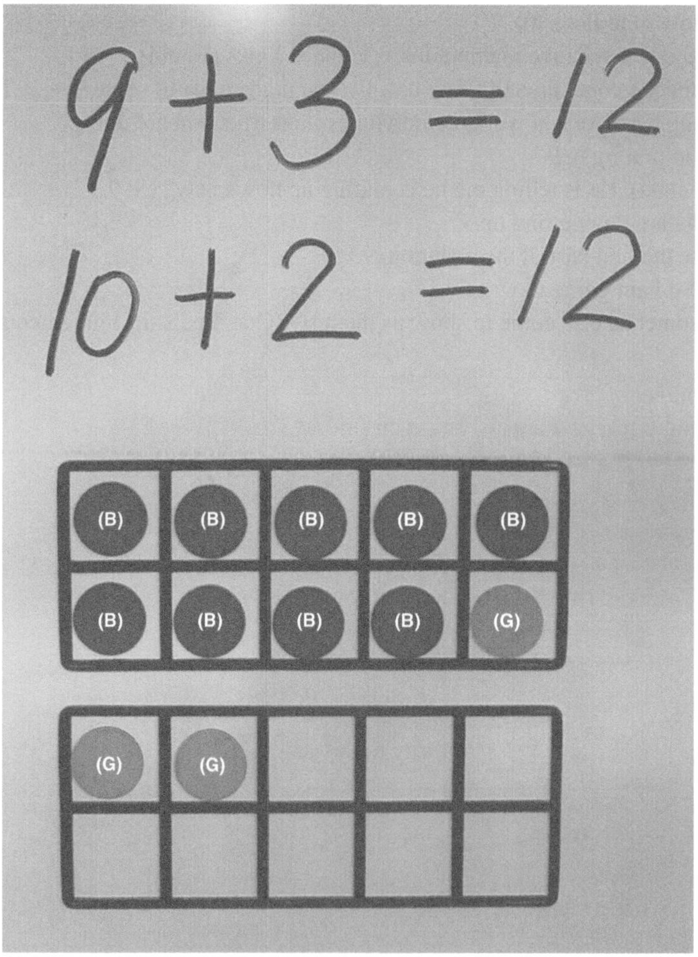

FIGURE 5.11 Teacher Y's Board work for solving 9 + 3

3. Discuss why make a 10

T: So, boys and girls, some addition facts, like 9 + 3, can be found by making a ten and adding to it. I'm wondering, why do we say "make a ten"? Why do you think we use 10? Why don't we use 5? 5s are good to work with.

S: 10 is easier to add.

T: 10 is easier to add. We can count by tens easily, it is an easy number to work with. Are there any questions?

S: Why do we use just ten? As you said, there are other easy numbers to add with.

T: What would you recommend? Which numbers do you think are easy to add?

S: 5 is easy, too. Two 5s make ten.

T: Right, but we just used ten to solve 9 + 3. So, making a ten can help us solve other addition equations.

4. Address student confusion with the example task

S: I'm a little bit confused about when they move the counter.

T: Would you like me to go back? So, when they move this green counter, that was a little confusing to you? Let's have someone else share again what they did. I am going to move this green counter back to where it originally was. Remember, they have 9 here and 3 here. Then they added all of these together. They told us that 9 + 3 was 12. Got that? Now, this is the tricky part. You mentioned they moved this. Can someone show her again why we moved the green counter and what happens in our equation when we move it? It's not going to match this anymore; it's not going to say 9 + 3.

S: (One S came up to show the process again). You move it to the ten frame to make ten. You could erase this up here and make it a 2 and a 10 (see Figure 5.12).

T: Great! When he moved this up here, it's not 9 anymore. How many are there now? (S: 10.)

T: 10, and how many are down here? (S: 2.)

T: 2. And what is the sum? (S: 12.) Which is the same thing. We didn't take anything away, and we didn't add anything. We are using the same total number of counters; we just made a what? (S: 10.)

T: We made a 10, why did we make a 10?

S: It's easier.

T: Exactly! It's easier to add with 10s, right? Awesome job!

T: Now, let's try a few practice problems in the textbook.

(The textbook suggests using ten-frames and writing a new equation to solve each problem).

5. Address student difficulties in practices

T: (Walking around while students are working)

T: (To class) I keep coming around and noticing that no one is adding that second addition fact, like we did up here when we made a ten. I should see two addition facts: the first one and how you made a ten to solve the problem.

T: (Continuing to walk around and help students)

T: (To the class) Some of you are forgetting, and this is the most important part of the lesson, to show your second equation. So, if we started with 8 + 5, we know it equals 13. When we made a ten like we did up here, what do we have? (S: 10 + 3.)

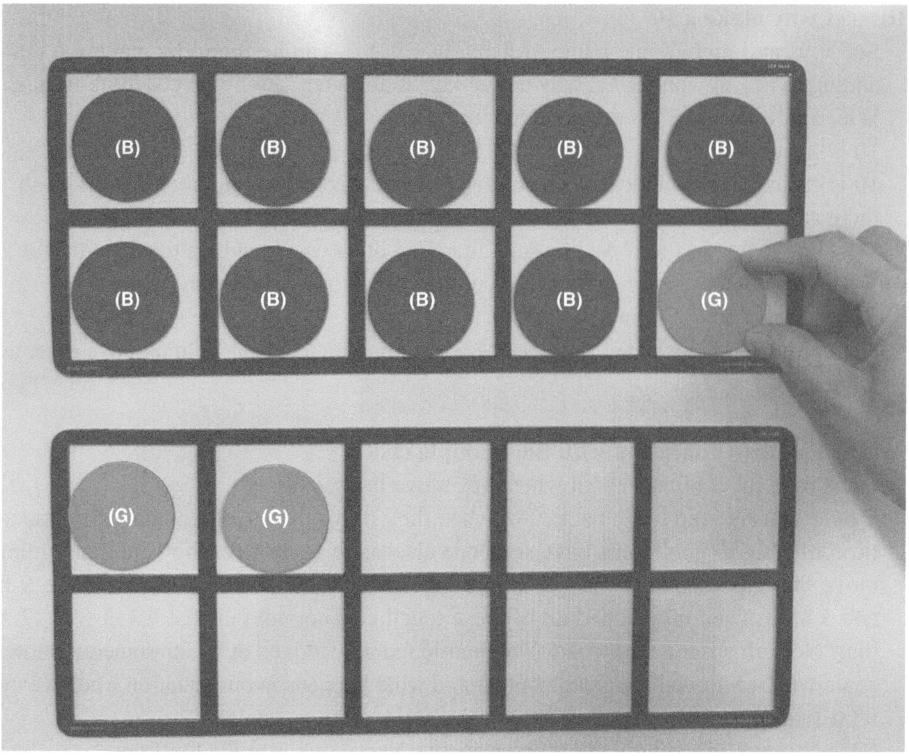

FIGURE 5.12 A student's board work redoing 9 + 3

T: And the sum will say the same… (S: 13.)

T: So I want to see both equations when you solve all the problems.
(Practice continued until the end of the class).

STEP 3: STOP AND REFLECT

Based on your reading and annotating Cases 7a and 7b, please reflect on the following:

1 **Representation uses.** Compare the representation uses of Teachers X and Y. How do their representational sequences differ? How are real-world contexts used differently?
2 **Deep Questioning.** Compare the deep questions asked by Teachers X and Y. How do they differ in terms of frequency and purpose of using deep questions?
3 **Linking to ICAP.** Based on the ICAP framework, which overall level of student engagement does each teacher likely promote? How might each teacher's lesson lead to the respective level?
4 **Insights Gained.** Compare your initial plan with the contrasting cases. What new insights have you gained that you can apply to your lesson planning and future teaching?

5.2.3 *Key to Analysis*

5.2.3.1 An Overview

The two sub-cases in Case 7 were adapted from two actual classroom lessons. While both lessons share similarities, they also have distinct focuses. Both aimed to teach the concept of making a 10 to add, which implicitly uses the associative property of addition—for example, $9 + 4 = 9 + (1 + 3) = (9 + 1) + 3 = 10 + 3$. This computation strategy aligns with the expectations of the Common Core. Additionally, both textbook lessons included a concrete real-world image, semi-concrete manipulatives (counters with a ten-frame), and corresponding number sentences.

However, the lesson foci differ. Case 7a emphasized sense-making of the "making a ten" process—specifically, what happens during the process of figuring out how many apples Mr. Monkey brought. This process was illustrated through both manipulatives and numerical recordings, highlighting the connections between the two. In contrast, Case 7b focused on computation. The real-world context (fireflies) that appeared in the textbook was not attended to by the teacher. Manipulatives (counters and two ten-frames) in this case were used to obtain computational results rather than to understand the process. Table 5.2 below highlights the key points and example deep questions from the two cases, followed by further elaborations.

5.2.3.2 Representation Uses

Teacher X in Case 7a began with a real-world context: apples brought by Mr. Monkey. Students were encouraged to observe the picture and pose a story problem. As discussed elsewhere, problem-posing provides an opportunity for deep learning. Through the context of Mr. Monkey's apples, students were guided to understand the make-a-ten-to-add strategy using counters and a ten-frame. Teacher X spent substantial time discussing the correspondence between what happened with the manipulatives and the related number sentence, helping students understand how and why one can turn 9 into 10 to add. Multiple strategies were illustrated with the manipulatives, and the effectiveness of each was discussed. Although this is a computational lesson, the real-world context supported meaning-making, and the overall sequence demonstrates concreteness fading.

In contrast, Teacher Y's lesson began directly with a computation task ($9 + 3$). Although the textbook included a picture of fireflies corresponding to $9 + 3$, the teacher overlooked it, missing the opportunity to situate the worked example within a real-world context. Manipulatives (two ten-frames with counters) were used to compute $9 + 3$, but the focus was on finding the answer. Rather than emphasizing understanding the process of "make-a-ten," Teacher Y guided students to record only the results of the manipulatives ($10 + 2$). As a result, students appeared confused about the make-a-ten process as they were unable to transform $9 + 3$ into $10 + 2$ on their own. Overall, the representational sequence in Case 7b contrasts with concreteness fading (abstract → semi-concrete → abstract) and remains computation-driven.

5.2.3.3 Deep Questions

The use of deep questioning in Teachers X and Y's lessons differed across three types: problem-posing, meaning-making, and comparison.

Problem-Posing Questions. As in other cases, Teacher X asks students to pose a question when a vivid picture—apples brought by Mr. Monkey—is presented: "Can you pose a question so we

TABLE 5.2 Key Points and Example Deep Questions in Case 7

	Case 7a	*Case 7b*
Representation uses	A real-world context → Counters → Number sentences. This sequence shows *concreteness fading*, with the word problem context being modeled to support sense-making of the make-a-ten-to-add strategy.	Numerical task → Manipulatives. This sequence contrasts with concreteness fading. The real-world context (firefly) was overlooked, and emphasis on the manipulatives' result rather than the process led to students' confusion.
Deep questioning	**Problem-Posing** • Can you pose a question so we can solve this problem using addition? **Meaning-Making** • Why did you use addition? • Can someone come to the board to explain to us how you figured it out how many total apples? • Why did you move 1 apple from these four into the box? • You can also separate 4 apples into 2 and 2. Why didn't you move 2 into this box? • What if we have enough space? Will you move 2 in? • Do we have other ways to compute this? • Who can explain this process completely? • Who can explain what this student said? **Comparison** • Compare these two ways, we are both making a 10 to add. Which number do you prefer to make a ten? 9 or 4? (9!) Why? • Let's compare this with what we did earlier (Referring to the first strategy). How are they different? • Which strategy do you prefer? • Which strategy did you like most? (make a 10 to add). How can we make a 10 to add? Can you talk to your desk mates?	**Problem-Posing** • NA **Meaning-Making** (*but all superficial responses*) • And why are you putting up 9? • Why did you put 3 up? • Why did you move this over there? • Why do you think we use 10? • Why don't we use 5? 5's are good to work with. • Can someone show her again why we moved the green one and what happens in our equation when we move it? **Comparisons** • NA
Linking to ICAP	Teach X's lesson likely promoted students' cognitive engagement to at least level C. The use of representations and questioning directed students' attention to the process of making a ten to add, going beyond the existing materials to teach the targeted method.	Teacher Y's lesson may have kept students' learning at level A. The use of representations and questioning focused students' attention on the results of manipulatives, which stayed at the level of existing materials and did not support understanding of the make-a-ten-to-add method.
Insights Gained	Answers may vary.	

can solve this problem using addition?" The proposed story problem—*There are 9 apples inside the box and 4 outside the box. How many apples are there in total?*—was then used as the worked example for the lesson. As analyzed elsewhere, using students' posed problems not only acknowledges their autonomy in knowledge construction but also motivates learning. More importantly, problem-posing facilitates students' inference-making and develops their creative thinking skills. Such problem-posing questions were not present in Case 7b.

Meaning-Making Questions. In both cases, Teacher X and Teacher Y asked a series of "why" questions that had the potential to promote deep learning. However, while Teacher X's questions elicited deep explanations about what happened during the "make-a-ten" process, Teacher Y's questions primarily directed students' attention to the results of the manipulatives rather than the process. As a result, many students in Teacher Y's lesson appeared confused about this strategy. This difference is evident in discussions involving the ten-frame manipulatives in the two classes.

In Case 7a, students used counters to represent apples. To compute $9 + 4$, a student moved 1 counter from outside the box into the box to make a ten, resulting in $10 + 3 = 13$. Teacher X asked, "*Why did you move 1 apple from these four into the box?*" Students responded, "This way, there will be 10 apples in the box. Ten plus something is easy!" This response indicates students' awareness of making a ten.

Although students mentioned the purpose of making a ten, Teacher X was not satisfied. The teacher asked a follow-up question to focus students' attention on the decomposing process: "*In your way, how did you separate your 4 apples? Which two parts did you come up with?*" This question highlighted that 1 was taken from 4 and that 4 was decomposed into 1 and 3. Teacher X then invited students to consider a counter-example: "*You can also separate 4 apples into 2 and 2. Why didn't you move 2 into this box?*" A student noted that there was not enough space in the box, which was true but missed the main point. Teacher X further prompted, "*What if we have enough space? Will you move 2 in?*" This redirected attention to the method: "*No, we want to make a 10 to add.*" These back-and-forth discussions helped students realize that the decomposing and composing process was purposeful. They laid a foundation for understanding the make-a-ten-to-add strategy as documented in its numerical expressions on the board (see Figure 5.8). Moreover, these discussions allowed students to make sense of the associative property of addition implicitly: $9 + 4 = 9 + (1 + 3) = (9 + 1) + 3 = 10 + 3 = 13$.

In Case 7b, Teacher Y also asked a few "why" questions but did not focus on the composing and decomposing process. Some questions were unnecessary. For instance, after students moved counters in the ten-frame for $9 + 3$, Teacher Y asked, "*And why are you putting up 9?*" and "*Why did you put 3 up?*" Although phrased as "why" questions, they did not demand deep thinking because they only reflected the literal transition from numbers to manipulatives. After students moved 1 counter from 3 to 9, Teacher Y asked, "*Why did you move this over there?*" Students indicated the purpose was to make a ten. While meaningful, unlike Teacher X, Teacher Y did not then guide attention to the composing/decomposing process.

Instead, Teacher Y asked, "*I'm wondering, why would I say make a ten? Why do you think we use 10? Why don't we use 5? 5's are good to work with.*" Students responded, "10 is easier to add," which Teacher Y confirmed: "10 is easier to add. We can count by tens easily; it is an easy number to work with." Interestingly, one student expressed curiosity and asked why 10 was used instead of 5, since the teacher had just mentioned that 5 was also easy. Teacher Y responded, "Right, but we just used a ten to solve $9 + 3$. So making a ten can help us solve other addition equations." This response did not address the student's concern.

In retrospect, Teacher Y could have emphasized the power of "10 + something" early in the review stage, as Teacher X did. During the new teaching, once students realized the importance of making a ten, Teacher Y should have moved further to orient students' attention to the composing/decomposing process, which was completely missing. As a result, many students remained confused. Teacher Y asked, "*Can someone show her again why we moved the green one and what happens in our equation when we move it?*" Although a good question, the follow-up simply repeated the previous explanation: after moving 1 to the ten-frame, 9 + 3 becomes 10 + 2. Students remained unclear about how 9 became 10 or how 3 became 2. Finally in the practice section, Teacher Y suggested using 8 + 5 = 13 to help arrive at 10 + 3, which appeared contrary to the intended purpose of the lesson.

Comparison Questions. In Case 7a, Teacher X created several opportunities for students to explore multiple solutions and compare different strategies. For example, after students computed 9 + 4 using two methods—either making 9 a 10 or making 4 a 10—Teacher X asked, "*Compare these two ways. We are both making a 10 to add. Which number do you prefer to make a ten, 9 or 4? Why?*" This question prompted students to reflect on the two solutions and recognize that making 9 a 10 is easier: "Because you only need to move 1 to make 9 a ten." This reflection allowed Teacher X to highlight a broader strategy: "So, we can first break the smaller number, then make the bigger number a 10." This differs from telling students an arbitrary rule that the smaller number must be made 10, which sometimes happens in classrooms. Moreover, when another student suggested breaking 4 into 2 and 2 to make 9 a 11, Teacher X documented this alternative strategy and invited the class to compare: "*Let's compare this with what we did earlier. How are they different? Which strategy do you prefer*?" Through comparing examples and counter-examples, students likely became more aware of the power of the make-a-ten-to-add strategy. Unfortunately, such opportunities for comparison were completely missing in Teacher Y's lesson in Case 7b.

5.2.3.4 Linking to ICAP

From the perspective of ICAP, the lesson in Case 7a likely promoted students' cognitive engagement to at least level C, whereas the lesson in Case 7b may have kept students' learning at level A. This difference arises from the distinct purposes of representation use in the two lessons, further reinforced by teachers' questioning.

As analyzed above, Teacher X's lesson used a real-world context involving apples, modeled with counters and a ten-frame. The purpose of the manipulatives was to support understanding of the decomposing and composing process, going beyond the existing materials. Teacher X's deep questions drew students' attention to this underlying process, helping them understand what occurs and make inferences during new learning. Therefore, it is reasonable to conclude that Teacher X's lesson promoted students' learning to at least level C.

In contrast, Teacher Y focused on guiding students through the manipulatives to perform the make-a-ten-to-add procedure and then document the resulting manipulations mathematically, while overlooking the existing firefly context. Because Teacher Y emphasized the results of the manipulatives, the questions did not prompt students to consider what happened during the make-a-ten-to-add process. Instead, the conversation was largely limited to describing existing/visible procedures and materials. Without opportunities for in-depth learning, students in this lesson struggled to generate new equations based on the manipulatives. In this sense, Teacher Y's lesson likely promoted students' learning only at level A.

5.2.3.5 *Insights Gained*

PSTs may have different takeaways from Case 7. As in other cases, PSTs in my methods course noted the use of concreteness fading and deep questioning in Teacher X's lesson (7a), compared with Teacher Y's lesson (7b). A typical response was:

> Some insights I have gained from this analysis include the importance of representation and deep questioning. Seeing examples of each teacher and the differences between them, along with the corresponding outcomes, I can see how important asking the right questions can be. Including these elements in my future lesson planning and teaching is very important to me after seeing these cases

Some PSTs were able to further explain why Teacher X's approach was more effective than Teacher Y's. For example, one PST noted that Teacher Y aimed to engage students, but this teacher's use of representations focused on the results, which did not support deep learning of the targeted make-a-ten-to-add methods.

> Teacher Y also attempted active engagement with the ten-frames and counters, having students come to the board to solve 9 + 3. However, this activity was not used to explore the "make-a-ten" strategy, as the focus remained on the results of the manipulatives. This led to confusion among students

As this PST analyzed, focusing on the results of manipulatives does not help students understand the underlying process, nor does it clarify the reasoning behind the corresponding numerical changes. Research (e.g., Moyer, 2001; Uttal et al., 1997) also indicates that although manipulatives can be useful in mathematics classrooms, their effectiveness depends on whether they are employed to support conceptual understanding. In this sense, focusing solely on the results of manipulatives contributes at most to obtaining the correct answer, but not to understanding the targeted method (e.g., how to make a ten to add), which represents the key shortcoming of Teacher Y's lesson.

5.3 Summary

In this chapter, I presented two pairs of Cases (6a and 6b, 7a and 7b) to illustrate how one might approach a worked example of the commutative or associative properties of addition, respectively. For the associative property, it was taught implicitly through the make-a-ten-to-add method. Each pair of cases demonstrated either the TEPS approach or a conventional approach.

The TEPS approach was illustrated by Teacher X's lessons. Across the two cases (6a and 7a), Teacher X situated the worked examples in real-world contexts (swimming pool; Mr. Monkey's apples). By presenting such pictorial contexts, students were likely motivated to learn. In both cases, the concrete situations were modeled with semi-concrete representations (e.g., ten-frame with counters), which led to numerical equations. This sequence demonstrates the concreteness-fading approach. More importantly, the real world context provided students with various reasoning opportunities, supporting inference-making and likely promoting learning up to the I and C levels. For instance, in Case 6a, the swimming pool context served as a platform that could be viewed in different ways. This generated multiple instances for comparison and pattern-seeking, likely encouraging students' critical and creative thinking skills. In Case 7a, the apple context provided a platform for students to understand the make-a-ten-to-add method, enabling them to grasp the hidden process beyond the visible results of manipulatives.

In both cases, Teacher X also posed various types of deep questions—including problem-posing, meaning-making, and comparisons. These prompted students to make sense of procedures and compare across instances, supporting inference-making about the targeted concepts (the commutative and associative properties). Teacher X's lessons likely promoted students' learning at the C or I levels of cognitive engagement.

In contrast, Teacher Y's lessons demonstrated a conventional approach. In both Cases 6b and 7b, Teacher Y began with either abstract definitions or a numerical task, followed by illustrations with semi-concrete tools (e.g., chips/blocks, counters with ten-frames). As such, the sequence went from abstract to concrete. Moreover, the basic properties were directly provided to students, leaving little room for sense-making. Although manipulatives were involved, Teacher Y did not ask deep questions about these representations. In this sense, Teacher Y's lessons likely maintained students' learning at Level A.

Piloting these cases in my methods course showed that PSTs generally grasped the TEPS approach. In particular, they were able to highlight differences in the use of representations and deep questioning, justified by the ICAP framework. When comparing their initial plans with the contrasting cases, PSTs expressed interest in incorporating insights from Teacher X's lessons—such as concreteness-fading that starts with real-world situations and deep questioning that elicits deep explanations—into their future planning and teaching.

References

Cai, J. (1998). An investigation of U.S. And Chinese students' mathematical problem posing and problem solving. *Mathematics Education Research Journal*, *10*(1), 3–15. https://doi.org/10.1007/BF03217121

Cai, J., Ran, H., Hwang, S., Ma, Y., Han, J., & Muirhead, F. (2023). Impact of prompts on students' mathematical problem posing. *The Journal of Mathematical Behavior*, *72*. https://doi.org/10.1016/j.jmathb.2023.101087

Cai, J., & Rott, B. (2024). On understanding mathematical problem-posing processes. *ZDM – Mathematics Education*, *56*(1), 61–71. https://doi.org/10.1007/s11858-023-01536-w

Chi, M. T., Adams, J., Bogusch, E. B., Bruchok, C., Kang, S., Lancaster, M., & Yaghmourian, D. L. (2018). Translating the ICAP theory of cognitive engagement into practice. *Cognitive Science*, *42*(6), 1777–1832. https://doi.org/10.1111/cogs.12626

Chi, M. T., & Wylie, R. (2014). The ICAP framework: Linking cognitive engagement to active learning outcomes. *Educational Psychologist*, *49*(4), 219–243. https://doi.org/10.1080/00461520.2014.965823

Moyer, P. S. (2001). *Are we having fun yet? How teachers use manipulatives to teach mathematics. Educational Studies in Mathematics*, *47*(2), 175–197.

Silver, E. A., & Cai, J. (1996). An analysis of arithmetic problem posing by middle school students. *Journal for Research in Mathematics Education*, *27*(5), 521–539. https://doi.org/10.5951/jresematheduc.27.5.0521

Sun, L., & Wang, N. (2012). *Elementary mathematics textbook (G1, v1)*. Jiang Su Phoenix Education Publishing.

Uttal, D. H., Scudder, K. V., & DeLoache, J. S. (1997). *Manipulatives as symbols: A new perspective on the use of concrete objects to teach mathematics. Journal of Applied Developmental Psychology*, *18*(1), 37–54.

6

TEACHING BASIC PROPERTIES OF MULTIPLICATION

Contrasting Cases 8–10

In this chapter, I present three cases (Cases 8, 9, and 10) on teaching the basic properties of multiplication: commutative, associative, and distributive[1]. Each case has two contrasting sub-cases employing different instructional approaches. In the TEPS approach (8a, 9a, and 10a), worked examples are situated in real-world contexts and modeled with semi-concrete representations (e.g., cubes, drawings) that gradually fade into abstract solutions. Each task is solved using two or multiple strategies, which were compared to illustrate the targeted concept. Deep questions promote problem-solving, meaning-making, and comparisons, supporting students' understanding of the targeted properties. In contrast, the conventional approach (8b, 9b, and 10b) usually begins with an abstract definition and numerical examples, followed by semi-concrete representations to verify them. Real-world contexts, when used, primarily serve as a pretext for computation. Teacher questioning focuses on obtaining correct answers, allowing any method, which limits opportunities for meaning-making. Lessons may contain uncorrected misconceptions, and comparisons, when opportunities arise, are often overlooked. As in earlier chapters, I analyze these instructional differences using the ICAP framework, highlighting their potential impact on students' cognitive engagement.

6.1 Teaching Commutative Property of Multiplication

6.1.1 Introduction

The Common Core State Standards expect Grade 3 students to apply the commutative property (CP) of multiplication in computation: "Apply properties of operations as strategies to multiply and divide. Examples: If $6 \times 4 = 24$ is known, then $4 \times 6 = 24$ is also known (Commutative property of multiplication) (3.OA.B.5)." To effectively use this property, students first need to understand its meaning, which is taught through the array model during multiplication lessons. In Case 8, we discuss teaching the CP of multiplication.

DOI: 10.4324/9781003586333-6

STEP 1: THINK AND PREPARE

Imagine you are a third-grade teacher. How would you teach CP of multiplication (e.g., $3 \times 6 = 6 \times 3$)? Draft a brief plan in a T-S dialogue format. What representations would you use, and how would you sequence them? What questions, including deep questions, would you ask, and what student explanations would you anticipate?

Your Initial Draft Plan:

6.1.2 Contrasting Cases: 8a and 8b

In Case 8, two subcases illustrate how third-grade teachers approach the worked examples of CP of multiplication (not full lessons). Case 8a features Teacher X using the TEPS approach, while Case 8b showcases Teacher Y employing a conventional method. Students in both classes have the same prior knowledge, and both lessons share the same teaching objectives:

Prior Knowledge. Students have already learned:

1 Equal groups model of multiplication
2 The basic meaning of multiplication

Teaching Objectives. Students will be able to:

1 Make sense of the CP of multiplication
2 Apply the CP of multiplication for computation

STEP 2: READ AND ANNOTATE

Below are cases 8a and 8b. Read and annotate each case, focusing on the use of representations and questioning. Specifically, for each teacher (X and Y), identify:

a the types of representations used
b the deep questions asked

Case 8a: Teacher X (TEPS)

1. Review the equal groups meaning

T: Our class is going to have a party! One parent brought 3 bags of apples, each with 6 apples (see Figure 6.1). Who can pose a question?

S: How many apples did this parent bring?

T: Who can state this story problem completely?

S: To prepare a party, one parent brought 3 bags of apples, each with 6 apples. How many apples did this parent bring?

FIGURE 6.1 Three bags of apples, with 6 in each

T: How many?
S: 18!
T: How did you figure that out? Which equation can we write? ($3 \times 6 = 18$).
T: Why did you use 3×6? (3 groups of 6).
T: Someone used 6×3 and got 18. Do you agree? Why? (Same answer, but it has a different meaning. 6×3 means 6 groups of 3, which isn't true here.)
T: Great! In this case, we should use 3×6 because we have 3 groups of 6.

2. New lesson: Pose a story problem
T: For our party, we will have a little play! I would like to arrange these 12 chairs in rows (see Figure 6.2), with the same number of chairs in each row.
T: Let's start thinking about all the possible ways we can arrange 12 chairs. How many rows could we have, and how many chairs in each row?
T: On your tables, each pair of you has 12 blocks, with each block representing one chair. Please work together to figure out all the possible arrangements.

3. Discuss student explore using manipulatives
(Students explore different possibilities for a few minutes in small groups)
T: What solutions did you come up with? I have 12 blocks up here on my smartboard screen. Can someone come to the board to show your solution? Explain what you have.
S: We have 3 rows, and each row has 4 chairs. (Move the blocks on the smart board. See Figure 6.3a.)
T: Great! This arrangement has rows and columns, and we called it an?
S: Array!
T: How many rows of chairs do we have? (3). And how many chairs are in each row? (4). This is called a 3-by-4 array. We have 3 rows and 4 columns.

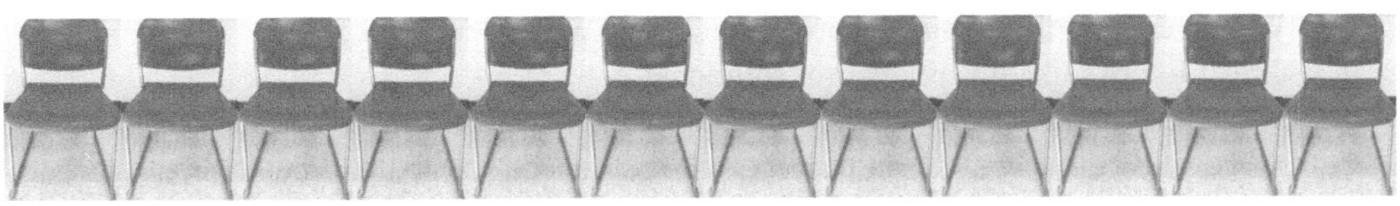

FIGURE 6.2 Arranging 12 chairs

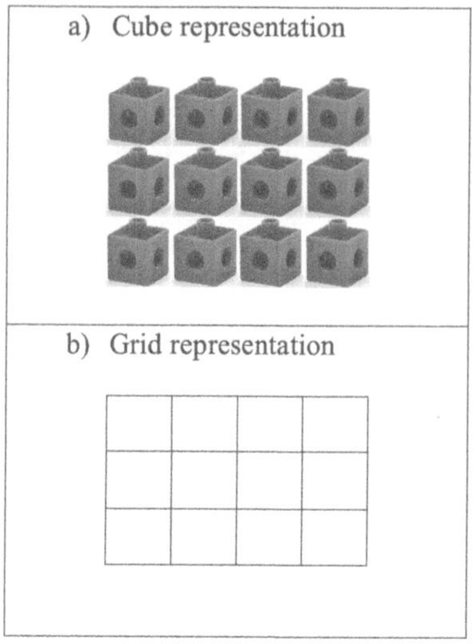

FIGURE 6.3 Arranging 12 chairs in a 3 × 4 array

T: How many of you had the same arrangement? Raise your hands.

T: Great! Let's draw this array on grid paper together. How many rows do we have here? (3). And how many chairs are in each row? (4). This is a 3-by-4 array on grid paper (see Figure 6.3b).

T: What equation does this arrangement give us? (S: $3 \times 4 = 12$).

T: Why 3×4? (S: 3 rows of 4 chairs).

T: Excellent. This arrangement matches $3 \times 4 = 12$.

T: Does anyone have an arrangement that turned sideways? Please come here to show us.

S: (Come to show 4 rows of 3 on the smart board) (see Figure 6.4a).

T: (Point to the blocks) How many rows do we have? (4). And how many chairs are in each row? (3). This is called a 4-by-3 array. We have 4 rows and 3 columns.

T: Now, let's draw a 4-by-3 array on your own grid paper. (Ask one S to draw on the board) (see Figure 6.4b).

T: What equation does this array give us? (S: $4 \times 3 = 12$, because we have 4 rows of 3).

T: Excellent! Let's take a look at these two arrays and two corresponding number sentences (see Figure 6.5).

T: Compare these two arrangements and equations. What do you notice?

S: (Possible responses: The two shapes are turned around. The two factors of these equations are switched, but the results, or products remain the same.)

T: What does this tell us about the relationship between 3×4 and 4×3? (They are the same.)

T: We can connect them using what symbol? (=). (Write: $3 \times 4 = 4 \times 3$)

T: Great! It looks like this is a pair of solutions!

T: What other arrangements did you come up with? If you have pairs of solutions like this, please draw them in this grid paper. (Invite students to come up, draw their arrays, and then share with the class).

S: We have 2 rows of 6 chairs and 6 rows of 2 chairs. We have $2 \times 6 = 12$ and $6 \times 2 = 12$.

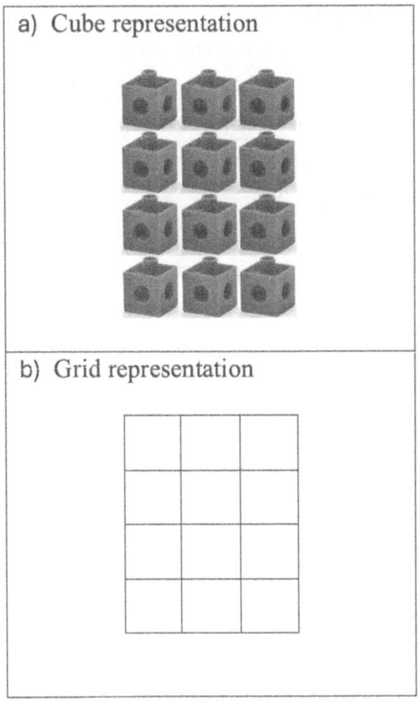

FIGURE 6.4 Arranging 12 chairs in a 4 × 3 array

T: Can someone help us understand how these two arrangements are different? (Students' explanation should mention "2 rows of 6 and 6 rows of 2").
T: What relationship do we see between 2 × 6 and 6 × 2?
S: They are equivalent.
T: So, we can connect them with what symbol? ("="). (Write: 2 × 6 = 6 × 2).
T: Any other type of arrangements?
S: We have 1 row of 12 and 12 rows of 1.
T: Interesting! Can you all imagine this type of chair arrangement? (Give students time to imagine).
T: What equations do we have? (1 × 12 = 12 and 12 × 1 = 12)
T: Comparing these two equations, what conclusion do you come up with?
S: 1 × 12 = 12 × 1. (T write on the board).

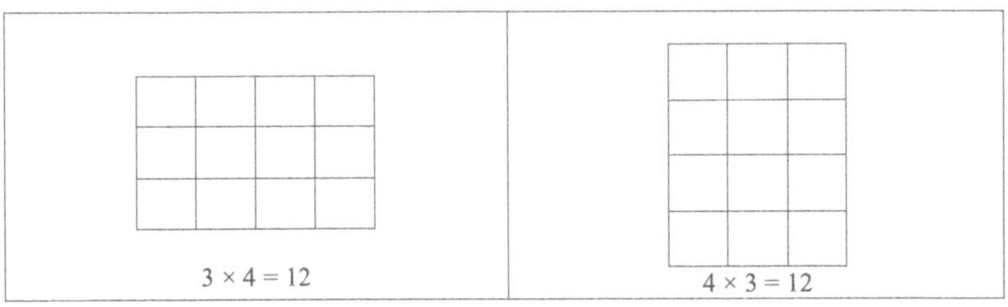

FIGURE 6.5 Comparing two different arrangements

4. Compare different pairs of equations

T: Wow! We have come up with so many different arrangements! Thank you for helping arrange our 12 chairs for our party! Let's take a look. What are the possible arrangements? (3-by-4 or 4-by-3 arrays; 2-by-6 or 6-by-2 arrays; 1 by 12 or 12 by 1 arrays).

T: Imagine you are sitting in the chairs for our play. Which arrangements do you like best? (Let students express their opinions)

T: We've also found pairs of equations based on these arrangements. Let's compare them — what do you notice? Take a moment to share your findings with your desk mates.

$1\times12=12\times1$

$2\times6=6\times2$

$3\times4=4\times3$

T: (After discussion) Now, would anyone like to share what you and your desk mates noticed? (Allow students to share their observations in their own words).

5. Reveal the big idea: The CP of multiplication

T: Excellent! When we multiply, we can switch the two numbers and still get the same answer. This is a big idea. We have learned something similar before. What was it?

S: The commutative property of addition!

T: Right! What may we call today's big idea?

S: The commutative property of multiplication!

T: Great! The commutative property of multiplication! How is knowing this big idea helpful?

S: (Possible responses:)

- If we know $4 \times 8 = 32$, we will know $8 \times 4 = 32$!
- We can switch two factors to check our multiplication.

T: Excellent! Today, we have learned the commutative property of multiplication- it's an important idea.

Case 8b: Teacher Y (Conventional)

1. Review the CP of addition

T: Hi, Friends! We've learned about turnaround facts, or what mathematicians call the commutative property of addition! Can someone give me an example? What does it mean? For instance, if we know $7 + 5 = 12$, what does that tell you?

S: It tells me $5 + 7 = 12$.

T: Can anyone give me another example?

S: $3 + 4 = 7$, so, $4 + 3 = 7$.

T: Great! What is the communtative property of addition?

S: You can switch two numbers when adding, and it will give you the same answer.

2. Introduce the CP of multiplication

T: Excellent. Today, we're going to learn about the same property, but with multiplication. Can you guess what we call it?

S: Turnaround facts with multiplication?

T: Correct. Mathematicians call it the commutative property of multiplication. So, when you multiply two numbers, you can switch the two numbers around and still get the same answer!

T: (Project the definition on the board). Let's read this statement together. "When multiplying two numbers, we can switch the two numbers and still get the same answer."

3. Give an example for computation

T: Let me give you an example: $3 \times 6 =?$ (18)

T: Correct. What does this tell you about $6 \times 3 =?$ (18)

T: So, 3×6 and 6×3 are the same! We can write $3 \times 6 = 6 \times 3$. Do you see how we switched the two numbers around? (Yes!)

T: This is an example of the commutative property of multiplication.

4. Illustrate the CP of multiplication with cubes

T: We can use cubes to prove that this is true.

T: I've passed each of you some cube sticks. In pairs, one of you will show 3×6, and the other will show 6×3. See if you end up with the same number of cubes.

(Students work in pairs.)

T: Who can share how you used your cubes?

S: In our group, I showed 3 groups of 6 (see Figure 6.6a). So, that's 3×6. My partner showed 6 groups of 3 cubes (see Figure 6.6b). So, that's 6×3. We both counted our cubes, and our totals were both 18.

T: Excellent! This shows that $3 \times 6 = 6 \times 3$.

T: Are there any other ways we could show this? Did you all use the same way? (Yes).

5. Illustrate the CP of multiplication with an array

T: (Show an array model). I did it slightly differently. Here are my cubes (see Figure 6.7a). I arranged them in rows and columns. What do we call this model? (Array!)

T: How many rows do you see? (3). How many columns? (6)

T: So, this is a 3-by-6 array, and our equation is 3×6.

T: Now, if I turn this array around, how many rows do you see? (6). How many columns? (3).

T: Yes, now it becomes a 6-by-3 array (see Figure 6.7b), and we can write the equation as 6×3.

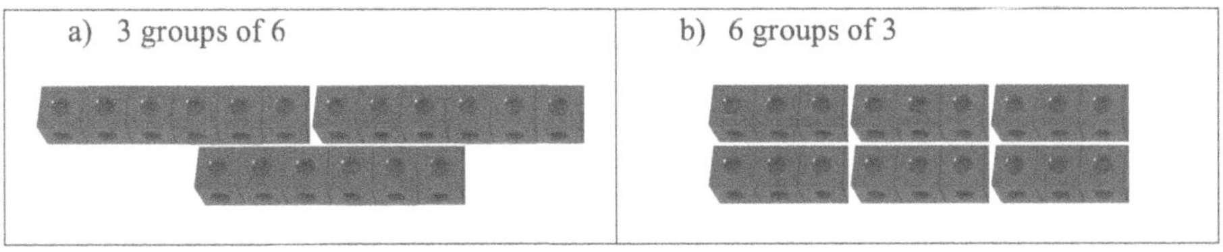

FIGURE 6.6 Students' representations of 3×6 and 6×3

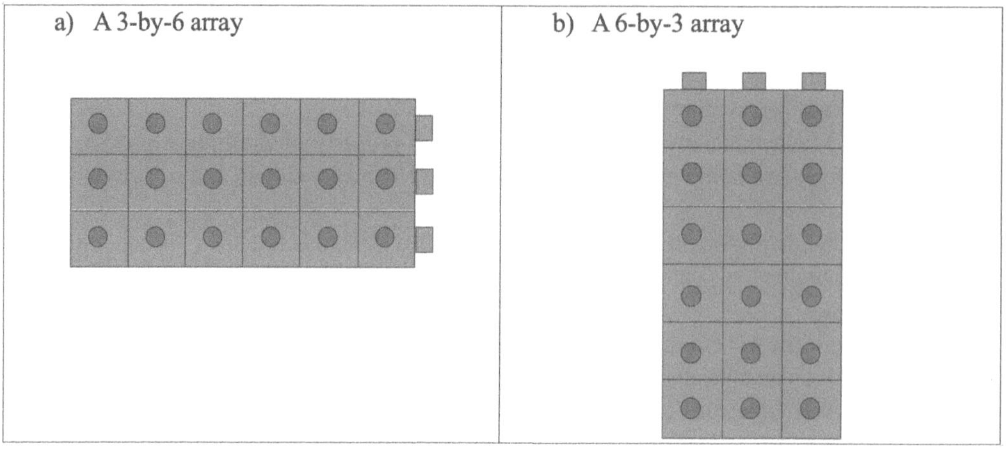

FIGURE 6.7 Teacher Y's representations of 3 × 6 and 6 × 3

T: Since I used the same array and simply turned it around, this shows us that $3 \times 6 = 6 \times 3$.
T: Both yours and my ways show the commentative property of multiplication!

6. Application CP of multiplication to solve word problems

T: Great job! Knowing the CP of multiplication can help us solve problems! Let's try these word problems. You can use the CP of multiplication to help you find the answers.

1. The movie theater arranged the chairs in 8 rows with 9 chairs in each row. How many chairs does the movie theater have?
2. Mom planted broccoli in the yard. There were 7 rows with 6 plants in each row. How many broccolis did Mom plant?

(Students work in their seats)

T: Let's check! What did you find for the first problem?
S: I used both 8×9 and 9×8 to solve this problem. My answer is 72.
T: Great job! Does everyone agree? (Yes!)
T: What about the second problem?
S: I used both 7×6 and 6×7 to solve this problem. My answer is 42.
T: Excellent! You all have learned the commutative property of multiplication so well!

STEP 3: STOP AND REFLECT

Based on your reading and annotating Cases 8a and 8b, please reflect on the following:

1 **Representation uses.** Compare the representation uses of Teachers X and Y. How do their representational sequences differ? How are real-world contexts used differently?

2 **Deep Questioning.** Compare the deep questions asked by Teachers X and Y. How do they differ in terms of frequency and purpose of using deep questions?
3 **Linking to ICAP.** Based on the ICAP framework, which overall level of student engagement does each teacher likely promote? How might each teacher's lesson lead to the respective level?
4 **Insights Gained.** Compare your initial plan with the contrasting cases. What new insights have you gained that you can apply to your lesson planning and future teaching?

6.1.3 Key to Analysis of Case 8

6.1.3.1 An Overview

The two lessons in Case 8 share some similarities. Both aim to teach the CP of multiplication and involve concrete story problem contexts, semi-concrete cubes, and abstract number sentences. However, the representational sequence and lesson focus differ. Case 8a emphasizes sense-making of the CP through a real-world context, whereas Case 8b presents the property directly. Table 6.1 presents key points and example deep questions, followed by elaborations.

6.1.3.2 Representation Uses

In Case 8a, the review task began with a real-world context (three bags of apples) to revisit the meaning of multiplication. The new lesson was situated in a story context of arranging chairs for a party. Students worked in small groups to create their own arrangements with cubes. Next, they were guided to draw arrays on grid paper. Both representations were compared and then connected to abstract number sentences, with an emphasis on the meaning of multiplication. This sequence illustrates *concreteness fading*. Notably, the same real-world context (arranging chairs) was viewed in different ways to generate different pairs of equations, offering students opportunities for pattern seeking. Based on these activities, Teacher X revealed the CP at the end, indicating a sequence that moved "from specific to general."

In Case 8b, Teacher Y began the lesson with the abstract definition of the CP. In other words, the property was given to students, which contrasts with Teacher X's lesson where students engaged in reasoning and discovery. The definition was then illustrated with a numerical example, followed by cube demonstrations in two arrangements. This sequence, moving from abstract to semi-concrete, differs from *concreteness fading*. Although the cube demonstration showed the CP in both equal groups and array models, it was not sufficiently used to support sense-making because the meaning of multiplication was not discussed. At the end, the teacher provided two story problems. However, both merely served as contexts for applying the CP. Without attention to the meaning of multiplication, students could be left confused about why both 8×9 and 9×8 worked.

6.1.3.3 Deep Questioning

As in other cases, the use of deep questioning in Teachers X and Y's lessons differed across three types: problem-posing, meaning-making, and comparison.

TABLE 6.1 Key Points and Example Deep Questions in Case 8

	Case 8a	*Case 8b*
Representation uses	A real-world context → Cubes → Drawing (arrays) → Number sentences. This sequence shows *concreteness fading*. The same real-world context is also viewed in different ways to generate different pairs of equations, providing opportunities for students to engage in pattern seeking. This sequence progresses from specific to general.	Definition → A numerical example → Cubes→ Real-world contexts. This sequence moves from abstract to semi-concrete and then to concrete, which is contrary to concreteness fading. The property is given to students. The counter illustrations are mainly used for demonstration, not for sense-making. The two-story problems in the end only served as a pretext for application of the CP.
Deep questioning	**Problem-Posing** • (After presenting picture). Who can pose a question? …Who can state this story problem completely?	**Problem-Posing** • NA
	Meaning-Making *In the review:* • Why did you use 3 × 6? (3 groups of 6). • Someone used 6 × 3 and got 18. Do you agree? Why? *In the new lesson:* • What equation does this give to us? (3 × 4 = 12). Why 3 × 4? (3 rows of 4 chairs).	**Meaning-Making** • NA
	Comparisons • Compare these two arrangements and equations. What do you find? • Can someone help us understand how these two arrangements are different? • Comparing these two equations, what conclusion do you come up with? • We've also found pairs of equations based on these arrangements. Let's compare them— what do you notice? • We have learned something similar before. What was it?	**Comparisons** • NA

Problem-Posing Question. In Case 8a, Teacher X asked students to pose a question during the review task. After presenting three bags of six apples, Teacher X asked, "Who can pose a question? … Who can state this story problem completely?" As analyzed earlier, consistent exposure to problem-posing opportunities is beneficial for students' deep learning (Cai & Rott, 2024; Cai et al., 2023). However, such an opportunity did not occur in Teacher Y's lesson in Case 8b.

Meaning-Making Question. Teacher X asked a set of meaning-making questions, which is completely absent from teacher Y's lesson. During the review task, when students suggested using

3 × 6 to represent "3 bags of 6 apples," Teacher X asked, "Why did you use 3 × 6?" This teacher also raised a common mistake, "Someone used 6 × 3 and got 18. Do you agree? Why?" Both questions emphasized the basic meaning of multiplication, laying a foundation for students' reasoning in the follow-up lesson. In the new lesson with the array model, Teacher X asked similar questions. For instance, with "three rows of four chairs," Teacher X prompted, "What equation does this give us?" "Why 3 × 4?" These questions encouraged students to make sense of the number sentences by referring back to the concrete and semi-concrete representations. However, Teacher Y did not ask such questions in Case 8b.

Comparison Question. In Case 8a, Teacher X asked a series of comparison questions. Students were given 12 cubes to make arrays on their own, which generated related arrangements (e.g., 4-by-3 and 3-by-4, 2-by-6 and 6-by-2) and corresponding equations (e.g., 4 × 3 and 3 × 4, 2 × 6 and 6 × 2). Teacher X invited students to observe and compare these arrangements both within and across pairs to identify the big idea. For example, the teacher asked, "*Compare these two arrangements and equations. What do you find?*" Students were expected to notice that the two shapes are rotated and the factors are switched, yet the products remain the same.

Additionally, Teacher X asked students to interpret the differences based on context. After the discussion of the first pair of arrangements, the teacher prompted students to make further inferences to discover the CP: "*Comparing these two equations, what conclusion do you come up with?*" and "*We've also found pairs of equations based on these arrangements. Let's compare them—what do you notice?*" Finally, Teacher X helped students connect the CP of multiplication to that of addition: "*We have learned something similar before. What was it?*" Students responded that they had also learned the CP of addition. These rich opportunities for comparison were absent from Teacher Y's lesson in Case 8b.

Nevertheless, Teacher Y's lesson presented several opportunities that could have been used for comparison. For instance, instead of simply telling students the big idea—"*So, 3 × 6 and 6 × 3 are the same! We can write 3 × 6 = 6 × 3. Do you see how we switched the two numbers around?*"—Teacher Y could have invited students to compare the two expressions and identify the pattern themselves. Similarly, for the two cube demonstrations, the teacher could have asked students to compare the corresponding number sentences in terms of equal groups. Finally, when solving the word problems, Teacher Y could have asked students to compare both solutions and explain why each worked, based on the story context and the meaning of multiplication.

6.1.3.4 Linking to ICAP

From the perspective of ICAP, Teacher X's lesson likely engaged students at the C and I levels, whereas Teacher Y's lesson limited students' learning to the A level. Teacher X's lesson involved students in multiple aspects of inference-making through representation use and deep questioning. As analyzed above, the chair context allowed students to create different arrangements, providing opportunities for Teacher X to ask meaning-making and comparison questions. Through these questions, students were prompted to make inferences collaboratively with peers—for example, reasoning about the meaning behind number sentences, connecting concrete and abstract representations, and identifying patterns across specific cases—which extended beyond the existing materials. Thus, it is reasonable to conclude that the cognitive engagement level in Teacher X's lesson reached levels C and I.

In contrast, Teacher Y's lesson began with a passive introduction of the definition of the CP of multiplication. Although cubes were used, they did not facilitate meaning-making. As analyzed above, no deep questions were posed to promote inference-making, despite potential opportunities. Consequently, this lesson likely only reached level A of cognitive engagement.

6.1.3.5 Insights Gained

PSTs may have different takeaways from Case 8. In my methods course, PSTs compared their initial lesson plans with both 8a and 8b and offered insightful reflections. One group of PSTs, who had worked together on their initial lesson plan, reflected on their use of representations. They felt that their lesson was better than Teacher Y's but could be improved by incorporating the ideas from Case 8a, such as using concreteness fading and starting with a real-world situation.

> Our initial work does not follow the concrete-to-abstract teaching method the way Case 8a does. For future planning, it would be better to start with our real-world example at the beginning of the lesson rather than at the end the way we did. Our beginning, however, was better than 8b because we did not immediately use the abstract math equation, but rather began with modeling the array model for multiplication.

Some PSTs were able to explain why Teacher X's lesson was more effective than Teacher Y's, considering both representation use and deep questioning. They also expressed strong interest in applying the insights gained to their future teaching. One PST wrote:

> Teacher X's use of real-life examples, like bags of apples and chair arrangements, made abstract ideas more accessible. Meanwhile, Teacher Y primarily focused on numbers and manipulatives, offering fewer real-world connections. Concrete examples like these can help students grasp difficult concepts before moving to symbols or equations. Additionally, Teacher X's use of thought-provoking questions—such as 'Why 3 × 4?'—encouraged students to think critically and explore relationships independently, unlike Teacher Y's method of giving answers, which limited engagement. This comparison shows the importance of asking questions that require students to think deeply and make their own connections. (…) For future lesson planning, I will aim to use real-world representations, thoughtful questioning, active learning, and scaffolding to guide students from hands-on activities to more abstract concepts. These insights reinforce my goal as a future teacher to bring real-life examples and concrete objects into my classroom to make learning meaningful and engaging

6.2 Teaching Associative Property of Multiplication

6.2.1 Introduction

The Common Core State Standards expect Grade 3 students to apply the associative property (AP) of multiplication for computation. "Apply properties of operations as strategies to multiply and divide. *Examples: If (...) 3 × 5 × 2 can be found by 3 × 5 = 15, then 15 × 2 = 30, or by 5 × 2 =10, then 3 × 10 = 30. (Associative property of multiplication.)* (3.OA.B.5)." To effectively apply this property, students first need to make sense of it —for example, understanding why *(3 × 5) × 2 = 3 × (5 × 2)* makes sense—which is initially taught through an equal groups model. While a volume model can also illustrate this property, the concept of volume is not formally introduced until fifth grade. In Case 9, we will discuss teaching the AP of multiplication.

STEP 1: THINK AND PREPARE

Imagine you are a third-grade teacher. How would you teach the associative property of multiplication [e.g., $(3 \times 5) \times 2 = 3 \times (5 \times 2)$] through an equal groups model? Draft a brief plan in a T–S dialogue format. What representations would you use, and how would you sequence them? What questions, including deep questions, would you ask, and what student explanations would you anticipate?

Your Initial Draft Plan:

6.2.2 Contrasting Cases: 9a and 9b

In Case 9, two subcases illustrate how third-grade teachers approach the worked examples of AP of multiplication (not a full lesson). Case 9a features Teacher X using the TEPS approach, while Case 9b showcases Teacher Y using a conventional method. Both lessons include a word problem based on the equal groups model from a US third-grade textbook. Students in both classes have the same prior knowledge, and the lessons share the same teaching objectives.

Prior Knowledge. Students have already learned:

1 The equal groups model of multiplication
2 The basic meaning of multiplication

Teaching Objectives. Students will learn be able to:

1 Make sense of the AP of multiplication
2 Apply the AP of multiplication for computation

STEP 2: READ AND ANNOTATE

Below are Cases 9a and 9b. Read and annotate each case, focusing on the use of representations and questioning. Specifically, for each teacher (X and Y), identify:

a the types of representations used
b the deep questions asked

Case 9a: Teacher X (TEPS)

1. Review the basic meaning of multiplication

T: Friends! Do you like Mangoes? (Show a picture of mangoes. See Figure 6.8). (Yes). They are delicious, aren't they?

T: I bring you some mangoes today! (see Figure 6.9). How many are in each bowl? (3). And how many bowls do we have? (4). So, how many total mangoes did I bring in total? Can you write an equation to solve it?

FIGURE 6.8 Picture of delicious mangoes

(Students work on the problem)

T: Let's share your solutions with the class. What equations did you come up with?

S1: $3\times4=12$

S2: $4\times3=12$

T: So, we have two different equations, but the same answer.

(To the students who provided equations)

T: Please explain the meaning of your equations. (To S1) Why did you use 3×4? (To S2) Why did you use 4×3?

S1: I use 3×4 because we have 3 mangoes in each plate, and there are 4 plates.

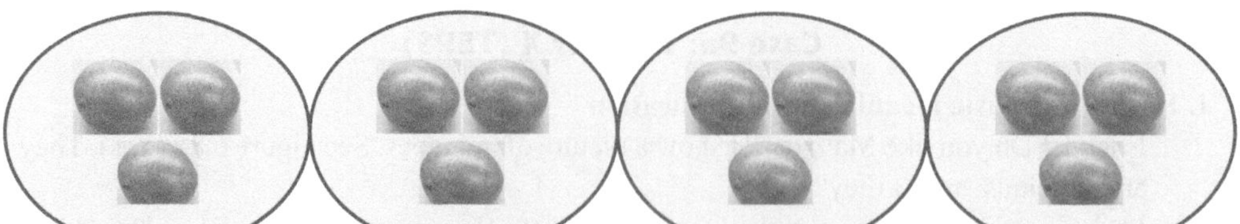

FIGURE 6.9 Four bowls of three mangoes

S2: I use 4×3 because we have 4 plates, and, there are 3 mangoes in each plate.
T: Let's take a closer look. How many groups do we have here? (Ss: 4). How many are in each group? (Ss: 3).
T: So, we have 4 groups of 3. Which equation should we use? (Ss: $4 \times 3 = 12$).
T: Excellent. This follows the basic meaning of multiplication. According to the math convention in the U.S., "*a* groups of *b*" is represented as "$a \times b$."

2. Introduce and represent the mango problem
T: With this knowledge, let's look at a more complex story problem about mangoes! (show the word problem on the board). Who can read?

Mr. Levin's students are testing foods grown in rainforests. He put 5 pieces of mango on each plate and put 2 plates on each table. There are 3 tables. How many pieces of mango are there?

T: To figure out the total number of mango pieces, let's start with drawing a picture to show what we already know. You can do this in small groups.
(Students work together to draw pictures while the teacher works around to observe).
T: (Projects selected student drawings on the board). I noticed that we've drawn some similar pictures (see Figure 6.10).
T: Can someone explain your picture? (Let students explain).

3. Solve the mango problem with multiple solutions
T: Now, based on this picture, how can we find out the total pieces of mango? Please work individually to write equations that make sense based on this picture.
(Students work individually.)
T: What equation did you come up with?
S1: I used $6 \times 5 = 30$!
S2: I used $5 \times 6 = 30$!
S3: I have $3 \times 10 = 30$!
S4: I have $10 \times 3 = 30$!
S5: I came up with $5 \times 2 \times 3 = 30$!
S6: I came up with $2 \times 5 \times 3 = 30$!
S7: I came up with $3 \times 2 \times 5 = 30$!
T: Wow! It seems that we all agree there are 30 pieces of mango. We also have different equations. Let's take a look and see which equation best aligns with our story context.

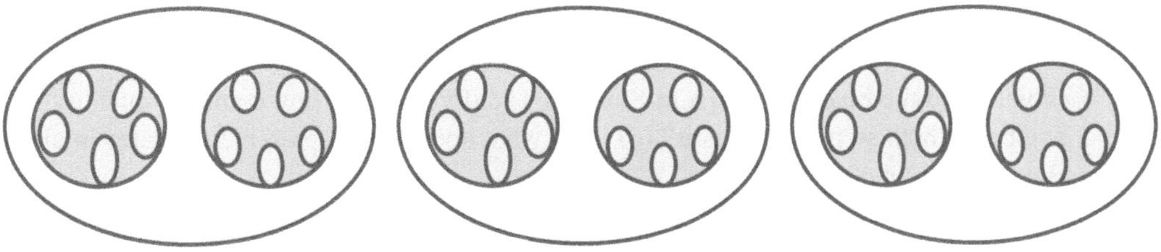

FIGURE 6.10 Typical student representation of the Mango problem

4. Discuss two solutions based on student input

T: Some students used $6 \times 5 = 30$ or $5 \times 6 = 30$. I am wondering where you got the 6 from. This story problem didn't mention 6, right?

S: There are 6 plates!

T: How do you know there are 6 plates?

S: There are 3 tables with 2 plates on each table.

T: Oh, right! What equation should we use to represent 3 tables of 2 plates?

S: $3 \times 2 = 6$.

T: (Write $3 \times 2 = 6$ on the board).

T: Why not 2×3?

S: We have 3 groups of 2 rather than 2 groups of 3.

T: Okay, so in this case, you first find out "How many plates are there?" (write on the board), which is…? (6!).

T: Now we know there are 6 plates. Each has how many mangoes? (5 mangoes). Can we find out "How many mangoes are there in total?" (write on the board).

S: $6 \times 5 = 30$

T: (Write $6 \times 5 = 30$ on the board.)

T: Anybody use $5 \times 6 = 30$? Why not?

S: We don't have 5 groups of 6.

T: Excellent. So, there are a total of 30 pieces of mango.

T: Some students used 3×10 and 10×3, which seems to be a different solution. In this solution, where did you get 10 from? There is no 10 in the original story context!

S: We found there are two plates with 5 mangoes each, which makes 10 mangoes on one table.

T: Great! So, you first find "how many mangoes are on one table" (write on the board).

T: How did you come up with 10? What equation did you use?

S: $2 \times 5 = 10$ because we have 2 groups of 5.

T: Why not "5×2"?

S: We don't have 5 groups of 2.

T: Great! (write $2 \times 5 = 10$ on the board).

T: Now, we know there are 10 mangoes on one table and we have 3 tables. How can we find the total number of mangoes? (write "How many mangoes are in total").

S: (May respond with both 3×10 and 10×3)

T: Let's take a closer look. How many groups of what do we have? (3 groups of 10, so 3×10)

T: Yes, we have 3 tables of 10, so, it will be $3 \times 10 = 30$. We have a total of 30 mangoes.

5. Combine two steps into one and compare two solutions

T: We just solved this problem with two different solutions, each using two steps. We can combine these two steps into one number sentence.

T: (Pointing to the board writing.) In Solution 1, did you notice any common quantity or number in both steps? What is it? (6). And what does the 6 represent? (6 plates in all). How did we get 6? (3×2)

T: Great! We can replace the "6" in Step 2 with "3×2" from Step 1. What does this give us?

S: (3 × 2) × 5 = 30 mangoes. (T writes it on the board).

T: Now, in the second solution, there's another common quantity. What is it? (10). That's right, there were 10 mangoes on each table. How did we get 10? (2 × 5). So, we can replace the "10" in Step 2 with "2 × 5" from Step 1. What does that give us?

S: 3 × (2 × 5) = 30 mangoes. (T writes it on the board).
(Board writing appears as follows:)

Solution 1	Solution 2
(1) How many plates are there?	(1) How many mangoes are on one table?
3 × 2 = 6 (plates)	2 × 5 = 10 (mangoes)
(2) How many mangoes are in total?	(2) How many mangoes are in total?
6 × 5 = 30 (mangoes)	3 × 10 = 30 (mangoes)
Or: (3 × 2) × 5 = 30 (mangoes)	Or: 3 × (2 × 5) = 30 (mangoes)

T: We just solved this mango problem with two different solutions, and both give us 30 mangoes. How are these two solutions different?

S: In one solution, we first find the total number of plates, and in the other, we first found the number of mangoes on one table.

T: Great! Even though the two solutions are different, they both solved the same problem. So, what does this tell us about (3 × 2) × 5 and 3 × (2 × 5)? Which symbol can we use to connect these two number sentences?

S: The equal sign. (3 × 2) × 5 = 3 × (2 × 5).

T: Excellent!

6. Reveal the AP of multiplication

T: Let's compare the two sides of this equation, (3 × 2) × 5 = 3 × (2 × 5), what do you notice?
Possible responses:

S1: They both have the same three numbers in the same order.
S2: On the left, the first two numbers are grouped together to multiply, while on the right, the last two numbers are grouped first. The answer stays the same.

T: Good observations! This example shows us that when we multiply three numbers, we can first multiply the first two numbers and then multiply the result by the third number. Or, we can first multiply the last two numbers and then multiply the result by the first number, and the answer remains the same.

T: We learned a similar big idea in addition. Do you remember what it's called? (S: Associative property of addition!)

T: Correct! So, what might we call this big idea that we learned from today's lesson?

S: The associative property of multiplication!

T: Exactly! The mango problem, which we solved in two ways, helps illustrate why the associative property of multiplication makes sense. Great job!

S: $5 \times 2 \times 3$
T: Raise your hands if you agree! (Ss raise hands)
T: How do you know to multiply them together?
S: It asks for the total number of mangoes, and it gives us three numbers. We just need to multiply them together!
T: Excellent! What does 5, 2, and 3 mean?
S: 5 means "5 pieces of mango," 2 means "2 plates," and 3 means "3 tables."
T: Now, at your seats, can you work with your partners to think of two ways to find the answer? Remember the associative property of multiplication allows us to compute in different ways.
(S work with peers)
T: Who can share what you did?
S: One way is to group 5×2 first. That gives us 10. Then, $10 \times 3 = 30$. So, we have 30 mangoes. The second way is to group 2×3 first. That gives us 6. Then, $6 \times 5 = 30$. We also get 30 mangoes.
T: Who else used the same method? (Ss raise hands).
T: Any other different ways?
S: We can also multiply 3×5 to get 15, and then multiply 15 by 2. That also gives you 30.
T: Correct! We can multiply the three numbers in whatever order we like. The associative property of multiplication allows us to find the answer in different ways. Well done, Class!

STEP 3: STOP AND REFLECT

Based on your reading and annotating Cases 9a and 9b, please reflect on the following:

1. **Representation uses.** Compare the representation uses of Teachers X and Y. How do their representational sequences differ? How are real-world contexts used differently?
2. **Deep Questioning.** Compare the deep questions asked by Teachers X and Y. How do they differ in terms of frequency and purpose of using deep questions?
3. **Linking to ICAP.** Based on the ICAP framework, which overall level of student engagement does each teacher likely promote? How might each teacher's lesson lead to the respective level?
4. **Insights Gained.** Compare your initial plan with the contrasting cases. What new insights have you gained that you can apply to your lesson planning and future teaching?

6.2.3 Key to Analysis

6.2.3.1 An Overview

The two lessons in Case 9 share several similarities. Both aim to teach the AP of multiplication and use a concrete story problem, semi-concrete cubes, and abstract number sentences. However, the lessons differ in focus. Teacher X in Case 9a emphasizes sense-making of the AP through the problem-solving process, introducing the term "*associative property of multiplication*" at the end. In contrast, in Case 9b, Teacher Y presents the definition of the property directly at the beginning. While some sense-making occurs with counters, the basic meaning of multiplication is applied inconsistently, which may lead to confusion. Additionally, Case 9b focuses primarily on computation, with the real-world context (the Mango problem) serving more as a pretext. Table 6.2 below lists the key points and example deep questions, followed by elaborations.

TABLE 6.2 Key Points and Example Deep Questions for Cases 9a and 9b

	Case 9a	*Case 9b*
Representation uses	A real-world context → Semi-concrete sketches → Two numerical solutions → Definition of AP of multiplication. The above sequence (concrete → semi-concrete → abstract) illustrates *concreteness fading*, with the real-world context (the Mango problem) used to emphasize meaning-making and connections.	Definition of the AP → A numerical example→ Semi-concrete counters → A story problem. The above sequence (abstract → semi-concrete → concrete) is the opposite of *concreteness fading*. The real-world context (Mango problem) appears only at the end, serving as a pretext for computation, and meaning-making is not emphasized throughout the process.
Deep questioning	**Meaning-Making** • Please explain the meaning of your equations. (To S1) Why did you use 3 × 4? (To S2) Why did you use 4 × 3? • I am wondering where you got the 6 from? This story problem didn't mention 6, right? • Why not 2 × 3? (We have 3 groups of 2 rather than 2 groups of 3). • Anybody used 5 × 6 = 30? Why not? • In this solution, where did you get 10? There is no 10 in the original story context! • Why not "5 × 2"? (We don't have 5 groups of 2).	**Meaning-Making** • NA • (In fact, there were mistakes about the meaning of multiplication.)
	Comparisons • We just solved this Mango problem with two different solutions, and both give us 30 mangoes. How are these two solutions different? • So, what does this tell us about (3 × 2) × 5 and 3 × (2 × 5)? Which symbol can we use to connect these two number sentences? • Let's compare the two sides of this equation, (3 × 2) × 5 = 3 × (2 × 5), what do you notice?	**Comparisons** • Based on what we already know about the associative property of addition, can you guess what the associative property of multiplication might be?
Linking to ICAP	Teacher X's lesson likely prompted students' cognitive engagement up to level C, as it engaged them in inference-making through concreteness fading and deep questioning.	Teacher Y's lesson may have limited students' learning to level A, as the counters were not used to facilitate meaning-making. Additionally, mistakes occurred during the activities.
Insights Gained	Answers may vary.	

6.2.3.2 Representation Uses

In Case 9a, Teacher X began the review task with a real-world context, four bowls of three mangoes, to revisit the basic meaning of multiplication. This step is critical because understanding the AP of multiplication requires applying this meaning twice in a hierarchical manner. In the worked example, a two-step Mango word problem was represented using semi-concrete sketches, and students solved the problem in two different ways. Each step of the numerical sentences was grounded in the meaning of multiplication within this actual context. The two numerical solutions were then compared to reveal the big idea of the AP of multiplication. This sequence illustrates *concreteness fading*, and the processes of meaning-making likely promote students' deep learning.

In contrast, Teacher Y began with a review of the AP of addition in a numerical context. Based on this, the lesson directly introduced the definition of the AP of multiplication, illustrated with a numerical example—in other words, the AP of multiplication was presented to students passively. Semi-concrete counters were then used to demonstrate this property. However, the counter illustration was confusing because the meaning of multiplication was applied inconsistently. For instance, in Figure 6.11, the counter picture on the left does not show $(2 \times 3) \times 4$ but rather $4 \times (2 \times 3)$. This is because the pictures shows 4 groups of 6 rather than 6 groups of 4. Finally, the story problem about mangoes was provided, but, as in other cases described in this book, it functioned primarily as a pretext for computation. The generated equations could not be meaningfully connected back to the story context. Overall, the representational sequence is the opposite of *concreteness fading*, and the use of representations did not support meaning-making.

6.2.3.3 Deep Questions

Across Cases 9a and 9b, the use of two types of deep questions —meaning-making and comparison—differed in Teacher X's and Teacher Y's lessons. Note that Teacher X did not ask students to pose questions in this lesson, likely because the complexity of teaching the AP of multiplication left little room for such an activity.

Meaning-Making Questions. Early in the review task, when students solved the one-step word problem in different ways, Teacher X asked, "*Please explain the meaning of your equations. (To S1) Why did you use 3 × 4? (To S2) Why did you use 4 × 3?*" This discussion laid a foundation for the new lesson, in which the meaning of each step was emphasized. During the new lesson, Teacher X invited students to solve the Mango problem in their own ways, generating correct, incorrect, or incomplete equations. All of these solutions were used for further discussion, highlighting Teacher X's value of students' co-generation of knowledge. For instance, Teacher X asked, "*I am wondering where you got the 6 from? This story problem didn't mention 6, right?*" This question elicited students' explanation that there were six plates because there were "*3 tables with 2 plates on each table.*" After students arrived at the correct expression, 3×2, Teacher X drew attention to the incorrect alternative: "*Why not 2 × 3?*" This stressed the meaning of multiplication from another perspective. Prompts like these occurred throughout the discussion of the worked example. Students were expected to link numerical expressions to concrete representations and understand the underlying reasoning.

In contrast, Teacher Y did not ask questions to emphasize meaning-making in Case 9b. Although the teacher attempted to use counters to illustrate $(2 \times 3) \times 4 = 2 \times (3 \times 4)$, the meaning of multiplication was applied incorrectly. As mentioned earlier, the counter illustration on the left (see Figure 6.11) does not work appropriately. In fact, the teacher could have used the picture on the right (2 groups of 3 groups of 4), which can be interpreted in two ways to generate the equation above. Similarly, no deep questions were asked with the Mango problem; students were only expected to apply the property to find the answer. In this sense, the real-world context was not used to promote inference-making.

Comparison Questions. Teacher X asked several comparison questions. After students solved the mango problem in two different ways, the teacher prompted them to compare the solutions: "*We just solved this Mango problem with two different solutions, and both give us 30 mangoes. How are these two solutions different?*" Students were expected to notice differences in the reasoning process: "*In one solution, we first find the total number of plates, and in the other, we first found the number of mangoes on one table.*" Such reasoning remained grounded in the context of the problem.

Next, Teacher X directed students' attention to the connections between the two solutions: "*Great! Even though the two solutions are different, they both solved the same problem. So, what does this tell us about (3 × 2) × 5 and 3 × (2 × 5)? Which symbol can we use to connect these two number sentences?*" This question elicited an instance of the AP: $(3 \times 2) \times 5 = 3 \times (2 \times 5)$.

The teacher then asked students to move beyond the context and compare the two expressions abstractly: "*Let's compare the two sides of this equation, (3 × 2) × 5 = 3 × (2 × 5). What do you notice?*" The purpose of this comparison was to help students grasp the essence of the AP at a more abstract level. Overall, this series of comparison questions enabled students to make inferences and discover the underlying big idea—the AP of multiplication.

In contrast, in Case 9b, comparison questions were infrequent. At the beginning of the new lesson, Teacher Y asked students: "*Based on what we already know about the associative property of addition, can you guess what the associative property of multiplication might be?*" This question had the potential to help students connect the AP of addition and multiplication. However, since students had not yet learned the new property, they might not have been able to make meaningful references. As seen in their responses, students gave superficial guesses, such as "*It takes three numbers*" or "*You can choose any two numbers to multiply together.*" At the end of the lesson, a missed opportunity for comparison arose. For the Mango problem, one student applied the AP to find the answer in two ways, while another switched numbers using the CP. This could have highlighted the differences between AP and CP, but the teacher simply accepted all responses, missing the chance for deeper discussion.

6.2.3.4 Linking to ICAP

From the perspective of ICAP, Teacher X's lesson likely prompted students' cognitive engagement up to level C. In contrast, Teacher Y's lesson may have limited students' learning to level A. This is because Teacher X's lesson engaged students in inference-making through concreteness fading and deep questioning. As analyzed above, the mango context was first used for review and then incorporated into the worked example, providing students with space to make sense of the numerical solutions. Deep questioning (e.g., comparisons) was consistently used to help students seek connections, leading to the discovery of the associative property of multiplication. Additionally, students in this lesson were given the freedom to solve problems in their own ways. The student-generated solutions were used for further discussions, extending beyond the existing materials. As such, it is fair to say that the cognitive engagement level in Teacher X's lesson may have reached at least level C.

In contrast, Teacher Y's lesson in Case 9b started with a review of the AP of addition, followed by an introduction to the AP of multiplication, along with its formal definition. Although semi-concrete counters were used, they were not employed in a way that facilitated meaning-making; in fact, errors occurred. Moreover, very few deep questions were asked to promote inference-making. Although there was one comparison question, its purpose did not seem to encourage students to make deeper connections. Indeed, the elicited responses were superficial. As such, this lesson may have only reached level A of cognitive engagement.

6.2.3.5 Insights Gained

Among all the cases, Case 9, which focused on teaching the AP of Multiplication, appeared to be the most challenging for PSTs. This may be because this property involves a hierarchical structure of multiplication (groups of groups of objects) on each side of the equation, requiring fluency and flexibility in the basic understanding of multiplication. In Case 9, the Mango problem was selected from a U.S. third-grade textbook. In my prior research (Ding, 2016; Ding & Heffernan, 2018; Ding et al., 2013), this task was used to develop PSTs' learning to teach mathematics. It was found that while PSTs could learn from this suggested approach, the process included productive struggles, as it was quite different from what they had experienced in their prior K-12 education. The approach described in Case 9b was consistent with the presentation in the textbook.

When Case 9 was piloted through the Methods course, most PSTs expressed interest in Teacher X's situating of the lesson in a real-world context (the Mango problem), as in the earlier cases. A typical response was:

> I like how students are working together to create a picture. I also like how Teacher X included the use of a picture of a mango to make number sentences. Using concrete examples is something I will apply to my future lesson planning

However, some PSTs did not fully grasp the key differences in Teachers X and Y's approaches to the Mango problem. For example, one PST focused on *when* the Mango problem was presented:

> Both teachers allow time for students to work with their peers, but Teacher X allows for students to do so sooner. Teacher Y waits until the end of the lesson, which is great, but students could possibly learn from their peers earlier if they had a chance to collaborate

Nevertheless, as in other cases, PSTs recognized the importance of deep questioning and real-world connections. Below is a typical statement:

> The main insight we gained after analyzing both Teacher X's and Teacher Y's lessons is that it is important to have deep questions and to use real-world problems to help students understand. Teacher X used a real-world problem to help the students relate it to abstract ideas of the equation and written numbers. This teacher uses a lot of questioning to get the students thinking about what is the correct understanding, not just the procedure

Interestingly, some PSTs also commented on Teacher Y's attempt to relate the AP of addition to the AP of multiplication, which they found insightful. Despite this, they were rightly critical of Teacher Y's approach and expressed appreciation for how Teacher X conducted the lesson:

> One thing that Teacher Y did that Teacher X did not was connect the associative property of addition to the associative property of multiplication. We thought this was a good idea to build

> on what students already know about the associative property of addition. However, we did not like how the teacher introduced the idea of the associative property of multiplication. Overall, Teacher X did a great job and is a strong model. Teacher Y needs to include more deep questioning and allow students to collaborate to improve the lesson

Taking together, although Case 9 was challenging to fully grasp, PSTs were prompted to compare TEPS and conventional approaches and gained valuable insights. A final note: In both the Spring and Fall 2025 semesters, I allowed PSTs to choose either the CP of addition or the AP of multiplication for the contrasting case analysis assignment. The majority of the class avoided the latter, indicating a greater need to help PSTs overcome the fear of teaching the AP of multiplication.

6.3 Teaching Distributive Property of Multiplication Over Addition

6.3.1 Introduction

The Common Core State Standards expect Grade 3 students to apply the distributive property (DP) in computation. For example, students should understand that "*Knowing that* $8 \times 5 = 40$ *and* $8 \times 2 = 16$, *one can find* 8×7 *as* $8 \times (5 + 2) = (8 \times 5) + (8 \times 2) = 40 + 16 = 56$" (3.OA.B.5), which illustrates the DP. To apply this property effectively, students first need to make sense of it, typically introduced through multiplication lessons. In Case 10, we will discuss teaching the DP of multiplication over addition.

STEP 1: THINK AND PREPARE

Imagine you are a fourth-grade teacher. How would you teach the DP —e.g., $8 \times (5 + 2) = (8 \times 5) + (8 \times 2)$—through a multiplication lesson? Draft a brief plan in a T–S dialogue format. What representations would you use, and how would you sequence them? What questions, including deep questions, would you ask, and what student explanations would you anticipate?

Your Initial Draft Plan:

6.3.2 Contrasting Cases: 10a and 10b

In Case 10, two subcases illustrate how fourth-grade teachers approach the worked examples of DP (rather than a full lesson). Case 10a features Teacher X using the TEPS approach, while Case 10b showcases Teacher Y employing a conventional method. Students in both classes have the same prior knowledge, and the two lessons share the same teaching objectives:

Prior Knowledge. Students have already learned:

1 The equal groups model of multiplication
2 The basic meaning of multiplication

Teaching Objectives. Students will be able to:

1 Make sense of the DP
2 Apply the DP for computation

STEP 2: READ AND ANNOTATE

Below are cases 10a and 10b. Read and annotate each case, focusing on the use of representations and questioning. Specifically, for each teacher (X and Y), identify:

a the types of representations used
b the deep questions asked

Case 10a: Teacher X (TEPS)

1. Introduce the shopping context and guide students for modeling

T: Hi, students! Have you ever gone shopping for clothes with your parents?

S: Yes!

T: Ms. Smith went to a clothing store to shop for her class show. She picked out suits for the boys (see Figure 6.13). The price of a shirt is $14, and the price of a pair of pants is $16. There are 8 boys in her class. How much does she need to pay for the boy's clothes?

T: What information do we know? What do we need to find?

S: A shirt costs $14, and a pair of pants cost $16. There are 8 boys, so we need to buy 8 sets of clothes.

T: If I use one circle to represent one shirt and one triangle to represent one pair of pants (see Figure 6.14, most left). Can you work with your desk mate to represent all the shirts and pants Ms. Smith needs to buy?

(Ss work with peers; T walks around and identifies typical work.)

FIGURE 6.13 The suit Ms. Smith picked out for boys

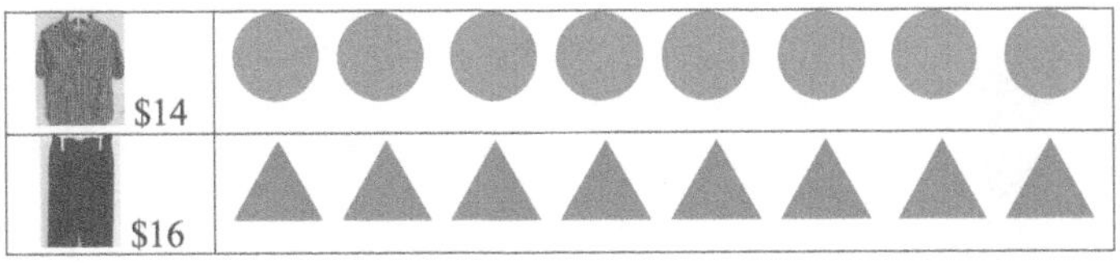

FIGURE 6.14 An example of student manipulatives

T: Who can come up to explain your group's work?
(S comes up to explain their manipulatives. See Figure 6.14)

S: We have 8 circles representing 8 shirts; and 8 triangles representing 8 pairs of pants.

T: Great! Now, with this information, can you work in small groups to figure out how much money Ms. Smith needs to spend? If you can think of different ways, please feel free to try them out.
(T walks around, observing and identifying different solutions. Students are likely to come up with two solutions).

2. Discuss two different solutions

T: Can someone come up to share your solution?

S: I first added \$14 and \$16, which equals to \$30. Then, I multiplied 30 by 8. I got \$240.

T: (Record it on the board). $14 + 16 = 30$, $30 \times 8 = 240$.

T: Can you explain what each step means? You can refer to this picture to explain.

S: $14 + 16 = 30$ means the cost of one set of clothes.

T: Can you highlight it in this picture? Come up here and show us.
(S comes to the board and circles a shirt and a pair of pants. See Figure 6.15).

T: Great. Your second step is 30×8. What does this mean? How many groups of what do we have? Do we have 30 groups of 8 here?

S: No, we don't have 30 groups of 8. We have 8 groups of \$30. I will change it to $8 \times 30 = 240$.

T: Oh, excellent! That makes sense!

T: This is a two-step word problem. How can we combine these two steps into one number sentence?

S: We can replace 30 in the second step with (14 + 16) from the first step.

T: Let's write it out:

$$8 \times (14+16)$$
$$= 8 \times 30$$
$$= 240\ (\$)$$

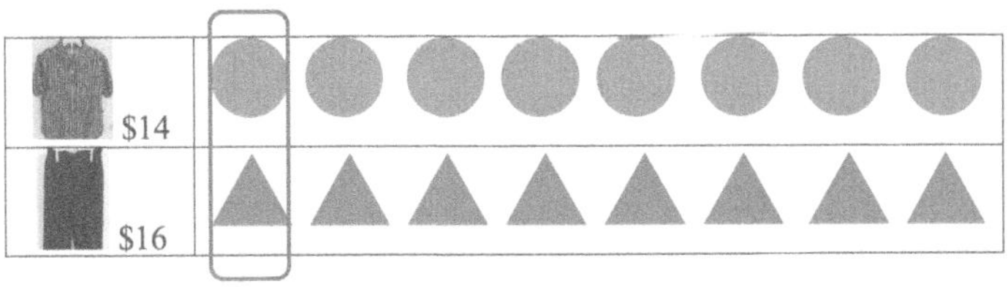

FIGURE 6.15 First solution illustrate with the representation

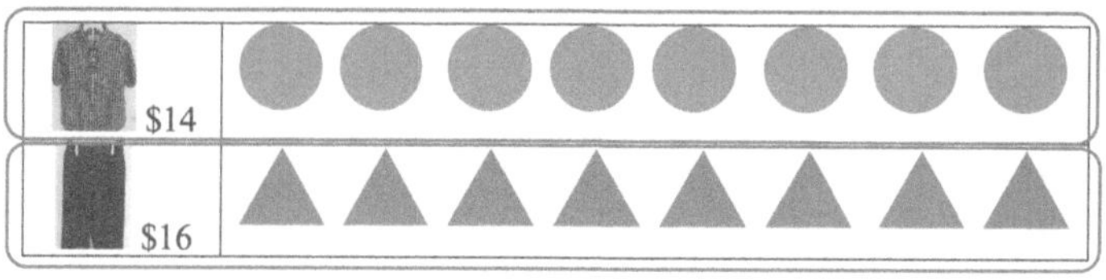

FIGURE 6.16 Second solution illustrate with the representation

T: Who has a different way to solve this problem?

S: We first figured out the cost of all shirts and then the cost of all the pants. We then added them together.

T: Excellent. Can you come up and write your solutions.

S: (writes): 8 × 14 = 112, 8 × 16 = 128, and 112 + 128 = 240.

T: Who can come up to explain your peer's solution using this picture?

S: They first find the total cost for all the shirts (circle the first row, see Figure 6.16 top) and then the total cost for all the pants (circle the second row, see Figure 6.16 bottom), and finally, they add them together for the total cost.

T: Great. We can also combine these steps into one number sentence. We can replace 112 and 124 with? (S: Replacing 112 with 8 × 14 and 128 with 8 × 16).

T: Exactly! So, we can write:

$$8\times14+8\times16$$
$$=112+128$$
$$=240\ (\$)$$

3. Compare the two solutions

T: Excellent job! We have solved this word problem using two different ways. Discuss with your desk mates how these two solutions, 8 × (14 + 16) and 8 × 14 + 8 × 16, are related. What sign should we use to connect these two solutions?

S: Since they both solved the same word problem, we can use an equal sign to connect them.

T: (Asks the class). Agree? (Yes). So, let's write: 8 × (14 + 16) = 8 × 14 + 8 × 16

T: Why are these two solutions equivalent? Can we explain both sides of this equation using the word problem context? Discuss this with your desk mates.

(Students discuss it with desk mates)

T: Who can share your group discussion?

S: In the first solution (on the left), we first find the cost for one set of clothes and then multiply that by 8 to get the total cost. In the second solution (on the right), we first find the cost for shirts and pants separately, and then we add them together to find the total cost.

T: Excellent! Now, compare the left and right sides of the equation. How are they the same, and how are they different? Let's focus on the numbers and operations only, without the context.

S: They have the same numbers.

S: On the left side, we multiply 8 by the sum of 14 and 16. On the right side, we multiply 8 by 14 and 16 separately, then add the two results. We find that the answer remains the same.

4. Try more examples and reveal the big idea

T: Very well-said! This is an important discovery! $8 \times (14 + 16) = 8 \times 14 + 8 \times 16$. That said, when we multiply a number by a sum of two numbers, we can multiply that number by each of the two numbers and then add the two partial products. This big idea is called the "distributive property."

T: What if we switch 8 and (14 + 16)? Can we still apply for this property? $(14 + 16) \times 8 =$?

S: Yes, $(14 + 16) \times 8 = 14 \times 8 + 16 \times 8$.

T: We can consider (14 + 16) groups of 8 to be the same as 14 groups of 8 plus 16 groups of 8. What if we switch the two sides of the equation? Does it still work?

T: Yes, $8 \times 14 + 8 \times 16 = 8 \times (14 + 16)$.

T: Great! This still shows the distributive property, but in the opposite direction. Here, we combined 14 and 16 to make a 30, and we factored 8 out.

T: Does the distributive property apply to other examples? Let's try a few. (Show three tasks). Please compute both sides to see if the results are the same.

1	$9 \times (11 + 89)$	$9 \times 11 + 9 \times 89$
2	$(100 + 2) \times 7$	$100 \times 7 + 2 \times 7$
3	$97 \times 11 + 3 \times 11$	$(97 + 3) \times 11$

Ss: (After computation, report the answers).

T: So, which sign can we use to connect both sides? (S: the "=")

T: Excellent! Despite the differences in each example, they all show the distributive property.

T: (Writes "Distributive Property" as the lesson title and asks students to read it together). This is what we learned from today's lesson.

T: Why do we call this big idea the "distributive property?" In the shopping example, which number is being distributed? (8).

T: Correct! (Pointing to board.) If you look at the equation from the left side to the right, you can see that we "distributed" 8 to both 14 and 16 to multiply them separately. However, if you look from the right side to the left, we factored out the same number, 8, and combined 14 and 16. This means we can use DP in both directions.

5. Sense the power of the distributive property

T: Why is it helpful to know the distributive property? Let's go back to our shopping example. Between the two solutions we discussed, which one was easier to compute?

S: Adding 14 and 16 first because it gives you 30.

T: Great. So, if we compute something like, $8 \times 14 + 8 \times 16$, instead of multiplying each part separately, we can add 14 and 16 first and then multiply by 8. That is, $8 \times 14 + 8 \times 16 = 8 \times (14 + 16)$.

T: What property allows us to do it this way? (S: The distributive property).

T: Let's look at another example on the board. When you compute $(100 + 2) \times 7$, Would you rather follow the order to compute 102×7 directly or to distribute the 7 to compute $100 \times 7 + 2 \times 7$? (S: The latter, as it is easier).

T: And what allows us to do it this way? (The distributive property).

T: Excellent! The distributive property is a powerful idea that makes our computation easier.

Case 10b: Teacher Y (Conventional)

1. Introduce the term the "Distributive Property"

T: Hi, students! Today we're going to learn about a really big idea. (Show the lesson objective). It says, "You will use the distributive property in order to multiply multi-digit numbers." Now, I'm reading that, and I see the word "distributive," a pretty big word, is underlined; but what does "distributive" mean? Think about all those vocabulary strategies you know, what do you notice in the word "distributive?"

S: It has the word "distribute."

T: Right! It has "distribute." What does "distribute" mean?

S: To break apart?

T: We're going break it apart; we're going to distribute things. <gesturing> Think about when teachers distribute the papers: we take a big stack of papers, and we distribute them to all the students. Okay? Does my number of papers change?

S: No.

T: No, it doesn't, it stays the same, it's just spread out, right? So, in math, to distribute, we have to think about our factors in multiplication, and we're going to break those factors apart. We'll distribute their value.

2. Teach the example of 3 × 17—Breaking apart 17

T: So, take a look at 3 times 17. I know my 3 times table; but, when I look at that 17, that's tricky. So, I'm going to break apart 17 because I can't do that in my head. So, what two numbers can we break 17 into that will make it easier to multiply? Write them on your board.

T: (to a student) Break your number 17 using place value. What plus what is 17?
(Students work on their small boards).

T: (to class) 1, 2, 3 and show me.
(students raise their boards)

T: (Taking two boards and holding them up) This one says we can break 17 into 10 and 7, and this one says 8 and 9. Raise your hand if you have the same idea.

T: Does anybody have another way we could break apart 17? So, we have 10 + 7 and 8 + 9, Any other ideas? (Writes 10 + 7 and 8 + 9 on the board)

T: What else could we break 17 into? What about if I start with 11?

S: Plus 6!

T: Plus 6, okay. (Writes 11 + 6)

T: So, we can break apart 17 in different ways, and any of these pairs will add up to 17.

3. Using an array model to illustrate the example

T: I can use a model to represent this multiplication. We can use an array or a grid. So, this grid has 1, 2, 3 rows, and each row has 17.

T: Today we're going to use this model to break down 17 into parts that are easier to multiply. (Pointing to the 17) I'm going to distribute this 17 into 2 groups. So, here we have 10 plus 7.

T: (Counting along the top from the left) 1, 2, 3, 4, 5, 6, 7, 8, 9, 10. (Outlines a 3-by-10 array) So here we have 3 rows of 10.

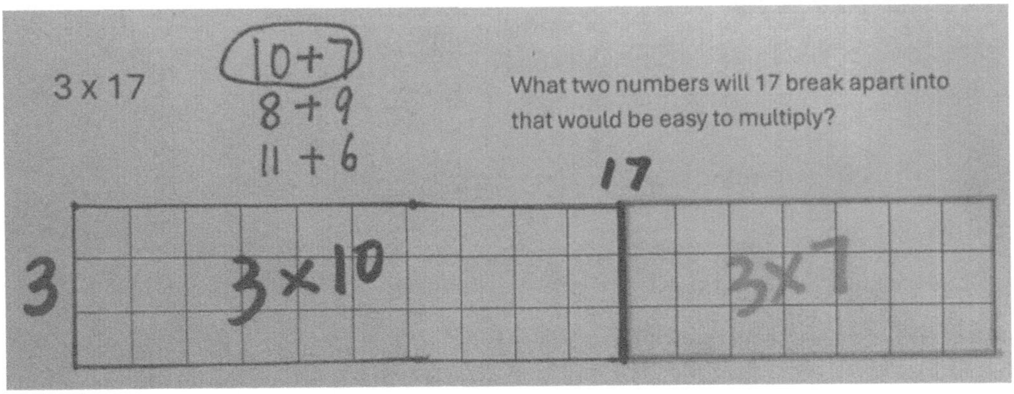

FIGURE 6.17 Using the array model to calculate 3 × 17

T: (Counts and circles the remaining part) Now, this part of the model will be our 7. Let count: 1, 2, 3, 4, 5, 6, 7. I still have 3 rows, but I have 7 in this part of the model (see Figure 6.17).
T: Did the size of my model change?
S: No.
T: No. Did the number of squares change?
S: No.
T: Correct. I just broke it up into different ways to make it easier to calculate. Now, on your board, what is 3 times 10? Write down your answer on your board.
T: Okay, show me! So, 3 times 10 is 30.
T: Now, on your board write 3 times 7. Ok, show me! Awesome. 21.
T: How am I going to get my total? What should I do with these two products?
S: Add them?
T: Add them!
S: 51
T: Good job, 51 (write 30 + 21 = 51). So, 3 times 17 equals 51.
T: I didn't have to multiply 3 by 17, but I broke 17 apart into two parts: 3 rows of 10 here, 3 rows of 7 there.
T: Today I want us to see if this method works all the time and to figure out the easiest way to break apart a number.

4. Students group work on 6 × 13

T: Take a look at that problem: 6 times 13. Which number do you think we should break apart to make it easier? On your board, write down which number you think we'll break apart.
T: Someone suggests 13. Raise your hand if you think it's going to be 6. Right, we're going to break apart the 2-digit number. So, let's take 13 and figure out how many ways we can break apart 13. Write down several possibilities on your board. You can work in small groups. Go!

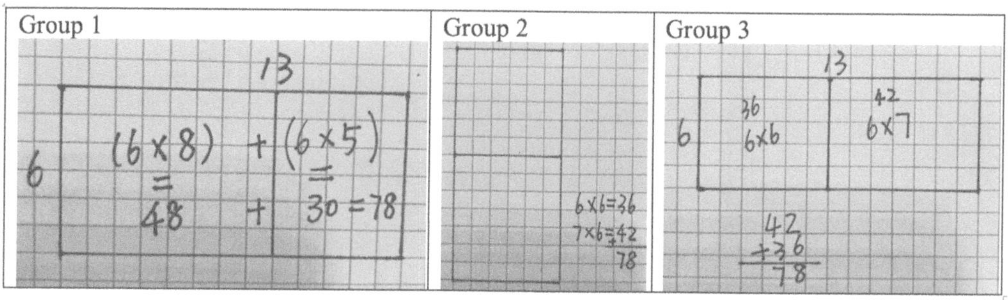

FIGURE 6.18 Students' group presentations on 6×13[2]

(After the group work. The teacher guided the class for report. Three groups presented. See Figure 6.18).

T: Alright, eyes on me. Now, here is the part where we look at our models and see what our product is. You are going to really look at what is the same and what is different. What do you notice?

T: Alright, Group 1, explain your model please?

S: So we knew that we had to break it up. We broke up 13, the big number, and we broke the 13 up into 5 and 8, because we know 8 plus 5 equals 13. So, the 6 stays the same, and we multiplied them, and for 6 times 5, we got 30; and for 6 times 8, we got 48. And then we added 48 plus 30 and we got 78.

T: Um, Group 2.

S: So what we did was, we put 6 squares on the bottom, and then we did 13 across, that's how we made our array. And then we decided to break 13 into 6 and 7, and then we counted how many were in each. And the first part was 36, and then we did the same to this one, which was 42. And then we added it up and we got the answer 78.

T: Okay. Go ahead, Group 3.

S: Um, we did, um, we splitted 13 into 6 plus 7, and then we did 6 times 6, which equals 36, and then we did 6 times 7 which equals 42. And then we did 36 plus 42 equals 78, and that's how we got our answer.

T: And your product is?

S: 78?

T: 78. What do you think about that? What do you notice about that?

S: It's the same?

T: It's the same. Hm.

5. Wrap up the lesson

T: Great job, Groups! How are these products all the same?

S: They have the same answer.

T: So what do you think their answers are going to be?

S: Probably they are going to be 78.

T: Probably they are going to be 78. This sounds like a rule to me. Something sounds like always true. We're going to talk about that next class.

STEP 3: STOP AND REFLECT

Based on your reading and annotating Cases 10a and 10b, please reflect on the following:

1 **Representation uses.** Compare the representation uses of Teachers X and Y. How do their representational sequences differ? How are real-world contexts used differently?
2 **Deep Questioning.** Compare the deep questions asked by Teachers X and Y. How do they differ in terms of frequency and purpose of using deep questions?
3 **Linking to ICAP.** Based on the ICAP framework, which overall level of student engagement does each teacher likely promote? How might each teacher's lesson lead to the respective level?
4 **Insights Gained.** Compare your initial plan with the contrasting cases. What new insights have you gained that you can apply to your lesson planning and future teaching?

6.3.3 *Key to Analysis of Case 10*

6.3.3.1 *An Overview*

The two lessons in Case 10 share similarities. Both aim to teach the concept of the distributive property (DP) and address its use in computation. However, the lessons differ in focus. Case 10a emphasizes sense-making of the DP through a real-world context, while Case 10b emphasizes computation, illustrated using an array model. Although both lessons address computation, Case 10a discusses both directions of using the DP— $(a+b)\ c = ac + bc$ (regular direction) and $ac + bc = (a + b)\ c$ (opposite direction, see Ding & Li, 2010) —whereas Case 10b primarily focuses on the regular direction, $(a+b)\ c = ac + bc$. Table 6.3 highlights detailed differences in representation uses and deep questioning.

6.3.3.2 *Representation Uses*

In Case 10a, Teacher X began with a real-world context—shopping for shirts and pants. The story problem was represented using semi-concrete counters organized in a table and then solved with abstract equations, demonstrating a sequence of concreteness fading. The problem was solved in two different ways, which were then compared to generate a new equation illustrating the DP. In this sense, the real-world context supports sense-making.

In contrast, Teacher Y began with an abstract definition of the DP, demonstrated through an example (3×17) that highlighted different ways to break apart 17. This was then illustrated with an array model, but no concrete context was provided. The array was decomposed in various ways to support computation; however, why thc DP makes sense and how it applies to computation were not explicitly addressed.

6.3.3.3 *Deep Questioning*

In Case 10, the deep questions in Teacher X's and Teacher Y's lessons differed mainly in two types: meaning-making and comparison, as elaborated below.

TABLE 6.3 Key Points and Example Deep Questions in Case 10

	Case 10a	*Case 10b*
Representation uses	A real-world context → Counters → Numerical equations. This sequence shows *concreteness fading*. The real-world context is used to support sense-making of the DP.	Abstract definition of DP → Numerical example (3 × 17) → Array. This sequence is different from concreteness fading. Real-world context is missing, and the array model is used for computation rather than sense-making.
Deep questioning	**Meaning/Connection Making** • Can you explain what each step means? You can refer to this picture to explain. • Your second step is 30 × 8. What does this mean? How many groups of what do we have? • Who can come up to explain your peer's solution using this picture? • Why do we call this big idea the "distributive property?" • Why do we need to know the distributive property? • What allows us to do it this way? (S: The distributive property).	**Meaning/Connection Making** • NA
	Comparisons • We have solved this word problem using two different ways. Discuss with your desk mates how these two solutions, 8 × (14 + 16) and 8 × 14 + 8 ×16, are related. What sign should we use to connect these two solutions? • Why are these two solutions equivalent? Can we explain both sides of this equation using the word problem context? Discuss this with your desk mates. • Now, compare the left and right sides of the equation. How are they the same, and how are they different? Let's focus on the numbers and operations only, without the context. • Between the two solutions we discussed, which one was easier to compute? (Variation) • What if we switch 8 and (14+16)? Can we apply the distributive property? • What if we switch the two sides of the equation? 8 × 14 + 8 × 16 = 8 × (14 + 16). Does it still work? • Who has a different way to solve this problem?	**Comparisons** • NA (*Questions with Potential*) • You are going to really look at what is the same, what is different. What do you notice? (*However, there is no explicit comparison made on the drawings.)* • How are these products all the same? (They have the same answer.) So what do you think their answers are going to be? (Probably going to be 78.). (*Only anticipated surface response.*) • Does anybody have another way we can break apart 17? *(The purpose was not on the targeted concept.)*

(*Continued*)

TABLE 6.3 (Continued)

	Case 10a	*Case 10b*
Linking to ICAP	Teacher X's lesson likely promoted students' cognitive engagement at the C or I level, as it engaged them in inference-making through concreteness fading and deep questioning. A real-world context (shopping) was used to situate the lesson, supporting students' sense-making of the targeted property. Multiple comparison opportunities were provided, allowing students to discover the DP.	Teacher Y's lesson likely limited students' learning to level A, as it focused on computation without opportunities for meaning making. The DP was presented to students primarily as a procedure for computation. Opportunities for comparisons were missed, and inaccuracy in the teacher's explanation of "distributive" likely cause student misconceptions.
Insights gained	Answers may vary.	

Meaning-Making Questions. In Case 10a, Teacher X asked a series of questions that encouraged students to explain the meaning of numerical expressions in relation to the concrete context. For example, after students proposed the first solution, the teacher asked, "*Can you explain what each step means? You can refer to this picture to explain.*" Later, following a second solution, Teacher X pressed further for understanding: "*Your second step is 30 × 8. What does this mean? How many groups of what do we have?*" This contributed to students' self-correction of a mistake. The teacher also invited students to explain each other's work: "*Who can come up to explain your peer's solution using this picture?*" These questions likely encouraged students to engage more actively with the ideas, co-generate new knowledge, make sense of them, and recognize connections among the different representations.

After the class had discussed the concrete worked example and explored multiple instances, Teacher X shifted to meaning-making of the mathematical term itself: "*Why do we call this big idea the 'distributive property'?*" Students were expected to recognize that one factor was distributed to two addends to be multiplied separately. To highlight the property's utility, the teacher also asked, "*Why do we need to know the distributive property?*" When the property was applied to simplify computation, Teacher X made its function explicit by asking, "*What allows us to do it this way?*" These questions focused students' attention on both the power and the purpose of the DP.

In contrast, Teacher Y in Case 10b did not ask meaning-making questions. Although the teacher did pose a question intended to clarify terminology, it risked leading to misconceptions. At the beginning of the lesson, Teacher Y drew students' attention to the word "*distributive*," highlighting its root "*distribute*." The teacher then asked, "*What does 'distribute' mean?*" A student responded, "*To break apart?*"—a guess the teacher accepted and reinforced: "*We're going to break it apart; we're going to distribute things. Think about when teachers distribute the papers.*" This brief exchange may have misled students into believing that the distributive property is only about breaking a number apart. Indeed, throughout the lesson, Teacher Y focused heavily on decomposing the number 17 (or 13) in different ways. It appeared that the teacher equated "distribute" with "break apart," stating, "*I'm going to distribute this 17 into two groups. So, here we have 10 plus 7.*" Consequently, when working with arrays, students concentrated on partitioning the array into parts. However, the essential idea—that one factor is distributed across two addends—was never made explicit.

Comparison Questions. In Case 10a, Teacher X asked a series of comparison questions to guide students' understanding of the distributive property. After the word problem was solved in two ways, the teacher prompted an explicit comparison: "*We have solved this word problem using two different ways. Discuss with your desk mates how these two solutions, 8 × (14 + 16) and*

8 × 14 + 8 × 16, are related. What sign should we use to connect these two solutions?" This encouraged students to focus on the relationships, leading to an instance of the distributive property: $8 \times (14 + 16) = 8 \times 14 + 8 \times 16$. Building on this, Teacher X pressed further: "*Why are these two solutions equivalent? Can we explain both sides of this equation using the word problem context? Discuss this with your desk mates.*" Students were expected to connect each side of the equation to the problem context and contrast the two approaches—one first calculating the total cost of an outfit, the other calculating the cost of shirts and pants separately. To promote abstraction, Teacher X shifted the focus: "*Now, compare the left and right sides of the equation. How are they the same, and how are they different? Let's focus on the numbers and operations only, without the context.*" The teacher also encouraged reflection on computational efficiency: "*Between the two solutions we discussed, which one was easier to compute?*"

Beyond direct comparisons, Teacher X posed "what if" questions to extend reasoning: "*What if we switch 8 and (14 + 16)? Can we apply the distributive property?*" and "*What if we switch the two sides of the equation? 8 × 14 + 8 × 16 = 8 × (14 + 16). Does it still work?*" Such variations encouraged students to test the boundaries of the property and explore its structural flexibility. Teacher X also invited alternative strategies: "*Who has a different way to solve this problem*?" This use of variation generated distinct but related instances that fostered meaningful comparisons and deepened students' understanding of the DP.

In contrast, Teacher Y in Case 10b did not pose explicit comparison questions, though opportunities were present. Before groups shared their work, Teacher Y told the class: "*You are going to really look at what is the same, what is different. What do you notice?*" Each group then presented how they computed 13×6 by decomposing 13 in different ways. Afterward, the teacher followed up with "*What do you notice?*" but this drew attention only to the identical answer, 78. No explicit comparisons were made across strategies to highlight how they were similar or different. At the end of the lesson, Teacher Y asked: "*How are these products all the same*?"—a question with potential for deeper comparison, but students again responded only that the answers were the same (78), a surface-level observation that did not engage with the structure of the expressions or the distributive property itself.

Likewise, in another task, Teacher Y asked: *"Does anybody have another way we can break apart 17?"* While this could have prompted useful comparisons, the teacher emphasized only additive decompositions of 17 (e.g., $10 + 7$, $8 + 9$, $11 + 6$) without examining how these decompositions supported different applications of the DP to 3×17. As a result, students' learning likely remained at a surface level—focused on breaking numbers apart rather than making inferences about the underlying mathematical structure.

6.3.3.4 *Linking to ICAP*

From the perspective of the ICAP framework, Teacher X in Case 10a used representations and deep questioning in ways that facilitated new inference-making, likely promoting students' cognitive engagement at the C or even I level. Students had the opportunity to discover the DP through problem-solving. The use of a real-world context helped them make sense of numerical expressions and draw meaningful connections across representations. Additionally, Teacher X posed a range of comparison questions that encouraged students to analyze both sides of a DP instance, with and without contextual support, and to compare across multiple instances. These comparisons required students to go beyond the given materials to grasp the essence of the property. In this sense, students in Teacher X's class were offered rich opportunities for deep learning.

In contrast, Teacher Y in Case 10b focused primarily on computation. Although the DP was the stated goal, its definition was introduced immediately at the outset and applied through repeated numerical decompositions, without giving students opportunities to explore, represent, or make sense of the property themselves. Unlike Teacher X's guided discovery, this approach resulted in passive delivery and limited conceptual engagement. Moreover, Teacher Y appeared to conflate distribution with simply breaking apart a number. This oversimplified presentation shaped the remainder of the lesson, as students were asked to decompose one factor in different ways—such as splitting 17 into 10 + 7 or 8 + 9—without attention to the DP's structural function. These tasks relied only on basic addition and subtraction and did not meaningfully expand fourth graders' mathematical understanding. Reducing the DP to number splitting may have left students with a limited or even incorrect understanding of the concept. Although students worked in small groups and created poster representations of their solutions, their cognitive engagement likely remained at the A level, without the reasoning or inference-making associated with deep learning.

6.3.3.5 Insights Gained

PSTs may have different takeaways from Case 10. In my methods course, a few PSTs thought that although Teachers X and Y's approaches differed, students in both classrooms demonstrated understanding. The majority, however, grasped the key differences between the two lessons, often drawing on the ICAP framework to justify their reasoning. For example, PSTs frequently highlighted Teacher X's use of representations—especially real-world connections—and the frequent use of deep questioning. Several PSTs also recognized that Teacher X guided students to solve problems in ways that naturally revealed the DP, rather than presenting it directly as Teacher Y did. In contrast, they noted that "*Teacher Y did not allow students to get much meaning out of the lesson.*" Some further pointed out Teacher Y's inappropriate explanation of *distribute*, noting that they would avoid such practice themselves:

> Teacher Y lacked deep questioning, proper representation, and correct terminology. This teacher translated the word distribute to 'break apart,' which does not represent the property because to distribute is to share parts of something. However, we still gained insights into what not to do as teachers. For this case, we would aim to be like Teacher X and not like Teacher Y

Beyond critiquing Teacher Y's lesson, one group of PSTs reflected on their own planning in light of the contrasting cases. They noted that they initially struggled with how to approach the task but appreciated seeing how Teacher X's lesson unfolded:

> When we were presented with the math equation $8 \times 14 + 8 \times 16 = 8 \times (14 + 16)$, we first would have taught based on the abstract problem. We considered making the connection between the two 8s. What do they mean? They both have 8 groups of something. But then, we struggled to understand our own teaching, and further, how our students would find meaning and understanding. I think we forgot about the use of representation and manipulation. Teacher X made a great scaffolded plan of concrete → abstract examples. They started with the price of clothes for an outfit and then represented that in circles and triangles, and finally in numerical examples. We really appreciated how they took the time to *represent* the distributive property in a real-life situation, then with models, and finally in abstract equations. In the future, we would aim to represent a lesson of this complexity in this way

6.4 Summary

In this chapter, I presented three pairs of cases (8a and 8b, 9a and 9b, 10a and 10b) to illustrate different approaches to teaching the CP, the AP, and the DP of multiplication[3]. The TEPS approach was illustrated through Teacher X's lessons, while the conventional approach was shown in Teacher Y's lessons.

As in earlier chapters, across the three TEPS cases (8a, 9a, and 10a), Teacher X situated worked examples in real-world contexts (e.g., arranging chairs, the Mango problem, shopping), which were then modeled with semi-concrete representations (e.g., sketches, counters, and shapes) and finally expressed as numerical equations serving as instances of the targeted properties. This sequence reflects a concreteness-fading approach. More importantly, the concrete and semi-concrete representations provided students with reasoning and inference-making opportunities. Across the three cases, the chair, mango, and shopping problems were each solved in two ways, which were then compared to generate an equation—an instance of the targeted property. In fact, in the chair problem, multiple arrangements for 12 chairs resulted in several instances of the CP of multiplication. Such problem-solving opportunities supported students' meaning-making of the targeted properties, likely promoting learning at the C or I levels of cognitive engagement.

Across all three TEPS lessons, Teacher X also posed deep questions, particularly those targeting meaning-making and comparisons. After guiding students to solve each task in multiple ways, Teacher X consistently asked them to compare the solutions, thereby generating corresponding instances. Students were then prompted to compare solutions with and without contextual support, helping them to grasp the targeted property in a more structural way. In this sense, Teacher X's lessons promoted students' learning beyond the existing materials, likely reaching the C or I level of cognitive engagement.

By contrast, the conventional approach to teaching the basic properties of multiplication (AP, CP, and DP) was illustrated through Teacher Y's lessons (Cases 8b, 9b, and 10b). In these lessons, Teacher Y directly introduced the definitions of the properties, followed by illustrations with numerical examples or semi-concrete tools. The focus remained on applying the properties for computation, with very few deep questions posed to support meaning-making or comparisons. Even in Case 10, although Teacher Y posed questions that could have facilitated comparisons, meaningful comparisons did not emerge, and only surface-level responses were anticipated. As a result, although students in Teacher Y's lessons were engaged in small-group work, their learning likely remained at Level A of cognitive engagement.

When these contrasting cases were used in my methods course with PSTs, I found that the AP of multiplication appeared to be the hardest for them to grasp. Nevertheless, across all three cases, PSTs generally recognized and appreciated the key elements of the TEPS approach. As in earlier cases, they expressed strong interest in deep questioning and in concreteness fading that begins with a real-world context. They were also able to justify their evaluations of the contrasting cases using the ICAP framework. Most also indicated that they would incorporate TEPS strategies into their future planning and teaching.

Notes

1. The distributive property of multiplication over addition, as its name suggests, is not purely a property of multiplication. For simplicity, I classify it as a property of multiplication, alongside the commutative and associative properties.
2. These representations are taken from an actual classroom. The images were redrawn by the author of the book.
3. The full name of the DP is the distributive property of multiplication over addition.

References

Cai, J., Ran, H., Hwang, S., Ma, Y., Han, J., & Muirhead, F. (2023). Impact of prompts on students' mathematical problem posing. *The Journal of Mathematical Behavior*, *72*. https://doi.org/10.1016/j.jmathb.2023.101087

Cai, J., & Rott, B. (2024). On understanding mathematical problem-posing processes. *ZDM – Mathematics Education*, *56*(1), 61–71. https://doi.org/10.1007/s11858-023-01536-w

Ding, M. (2016). Developing preservice elementary teachers' specialized content knowledge for teaching fundamental mathematical ideas: The case of associative property. *International Journal of STEM Education*, *3*(9), 1–19.

Ding, M., & Heffernan, K. (2018). Transferring specialized content knowledge to elementary classrooms: Preservice teachers' learning to teach the associative property. *International Journal of Mathematical Education in Science and Technology*, *49*, 899–921. https://doi.org/10.1080/0020739X.2018.1426793

Ding, M., Li, X., & Capraro, M. (2013). Preservice elementary teachers' knowledge for teaching the associative property: A preliminary analysis. *Journal of Mathematical Behavior*, *32*, 36–52. https://doi.org/10.1016/j.jmathb.2012.09.002

7

CONCLUSION

This chapter concludes the book by revisiting its overall goal and synthesizing insights from the contrasting cases in Chapters 3–6 through the lens of the *Teaching through Example-based Problem Solving* (TEPS) approach. By linking key features of TEPS—representation uses and deep questioning—to cross-case analyses, the chapter illustrates how teachers' instructional choices shape students' cognitive engagement, as framed by the ICAP model. Readers are reminded that all TEPS cases focus on teaching worked examples rather than full lessons, showing how problem-solving in real-world contexts can be unpacked to promote deep mathematical learning. Through guided comparisons, preservice teachers (PSTs) can observe how strategies such as concreteness fading and deep questioning foster inference-making, a hallmark of deep learning. While the book emphasizes early algebra, the TEPS approach is applicable to other domains of elementary mathematics. The chapter concludes by underscoring the value of contrasting cases and the ICAP framework in teacher education and by suggesting future directions, including leveraging technology and AI to further support PSTs' learning.

7.1 Purpose of This Book

The Association of Mathematics Teacher Educators (2017) has outlined standards for high-quality teacher preparation. In response, I have been working to develop systematic support for elementary teachers. My motivation originates from observing a persistent cycle in mathematics education that is difficult to break: Many elementary teachers in the U.S. have limited mathematical knowledge for teaching, which constrains their ability to plan and teach effectively. This, in turn, affects the next generation of teachers. Given that PSTs may serve as an entry point to break this cycle (Ma, 1999; Morris et al., 2009), I aim to leverage teacher education to prepare future elementary teachers to teach mathematics with depth.

As introduced at the beginning of this book, this work extends my prior NSF CAREER award (2014–2020), during which I collected hundreds of mathematical lessons from U.S. and Chinese expert teachers (Ding, 2021). I analyzed these lessons to glean insights from both contexts and synthesized an instructional approach called TEPS—*Teaching through Example-based Problem Solving*. To make TEPS more concrete and user-friendly for teacher education, I developed ten

DOI: 10.4324/9781003586333-7

pairs of contrasting cases, designed to serve as supplementary materials for mathematics methods courses. These cases can be analyzed and justified using the Interactive, Constructive, Active, and Passive (ICAP) framework (Chi & Wylie, 2014; Chi et al., 2018), which may, in turn, enrich ICAP from the perspective of mathematics education.

The targeted content in this book is early algebra, a critical domain in elementary school. Although various books on early algebra exist (e.g., Carpenter et al., 2003, 2015), most focus on documenting children's learning potential. As Carpenter et al. (2015) note, their goal was not to provide instructional models for teachers. While other books on teaching cases (Schifter et al., 2016a, 2016b) or instructional strategies (Fosnot & Jacob, 2010; Russell et al., 2011) exist, they neither facilitate direct comparison between cases nor systematically present a teaching approach like TEPS. In response, I developed contrasting cases of early algebra lessons to provide instructional support and to illustrate the TEPS approach in a structured and systematic way, thereby offering a distinctive perspective to the field.

Although primarily written for preservice elementary teachers learning to teach mathematics, this book is also suitable for in-service teachers seeking to improve their practice. Anyone involved in supporting elementary mathematics education—such as math coaches, professional development providers, methods course instructors, mathematics educators, researchers, and graduate students—may find this book helpful. Of course, TEPS is only one of many approaches that can effectively support elementary students' learning. Additionally, due to the diversity of actual classroom contexts, TEPS is not a panacea that fits all situations. Teachers who wish to employ this approach need to adapt it to their own classroom contexts.

7.2 Revisiting the TEPS Approach

As its name suggests, TEPS emphasizes teaching through example–based problem solving. The approach focuses on in-depth unpacking of a worked example through a problem-solving process. Importantly, TEPS is not positioned as a dichotomy between constructivist approaches (e.g., teaching through problem solving) and traditional methods (e.g., teaching through worked examples). Rather, it represents a balanced view that integrates insights from both. Similarly, the cases compared and contrasted do not reflect opposing philosophical camps; instead, they illustrate how a lesson can be taught differently, resulting in varying levels of cognitive engagement and, consequently, either deep or shallow learning. As shown in Chapters 3–6, the majority of cases focus on a single worked example rather than the full lesson. To unpack the worked example in-depth, TEPS emphasizes two instructional aspects: Representation uses and deep questioning.

Representation Uses. TEPS stresses concreteness fading, beginning with the most concrete, real-world situation. This situation is modeled with semi-concrete tools (e.g., counters, sketches) and then faded into numerical solutions. This concreteness fading sequence, Concrete–Semi-concrete–Abstract, has received considerable attention in the field (Ching & Wu, 2019; Fyfe et al., 2014, 2015; Ottmar & Landy, 2017) and is regarded as an instructional theory (Fyfe & Nathan, 2019). When teaching a worked example, the concreteness fading approach emphasizes coherence across representations: Each stage refers to the same situation, providing a platform for students to link and map between representations. This process promotes inference-making and is likely to enhance students' cognitive engagement up to the levels of C or I in the ICAP framework.

Note that TEPS strongly emphasizes situating the worked example in a real-world context, which connects to children's experiences and provides opportunities for problem posing and solving. In all ten cases, Teacher X began lessons with real-world pictures (e.g., apples, plums, swimming pools, shopping scenarios) to teach the targeted concepts. This is similar to, but distinct from, the CRA approach (Witzel, 2005), which follows the sequence concrete–representational–abstract. In CRA, "concrete" refers to hands-on materials (e.g., cubes) and "representational" refers to drawings or diagrams. In TEPS, the concrete stage is a real-world situation, represented by pictures, objects, or word problems. Emphasizing real-world connections is critical, as it enables teaching a worked example through problem solving and provides a platform for students to construct meaning from what they know, likely prompting deep cognitive engagement within the ICAP framework.

Deep Questioning. In the original TEPS model (Ding, 2021), I highlighted two types of deep questions: Concept-specific questions, which promote meaning-making, and comparison questions, which promote connection-making. These question types were found to significantly predict students' algebraic learning in elementary school, controlling for other factors (Ding et al., 2023a). In this book, I refer to them as "meaning-making" and "comparison" questions, respectively.

Meaning-making questions, as the name suggests, emphasize prompting students to make sense of mathematical ideas and concepts. For example, in Cases 1–10, Teacher X consistently asked students why they used a particular operation (e.g., addition, subtraction, multiplication, division) and why they chose a specific expression (e.g., 3×6 rather than 6×3). Such questions encourage students to seek mathematical meaning from the given situation, facilitating connections between concrete and abstract representations. They also support the inference of new knowledge beyond existing information, promoting cognitive engagement at least to the C level in the ICAP framework. Notably, meaning-making questions are less common in conventional classrooms (e.g., Teacher Y), where the focus is often on finding computational answers. As the National Research Council (2001) noted, U.S. mathematics instruction has traditionally emphasized procedures and rote memorization over conceptual understanding. This trend was also observed in my prior research with U.S. classrooms, including those of expert teachers (Ding & Carlson, 2013; Ding et al., 2019, 2021, 2023a, 2023b).

Comparison questions are common in the TEPS approach. Across all cases, Teacher X asked students to compare different operations to identify inverse relations (Cases 1–5) or different solutions to generate instances of the basic properties (Cases 6–10). According to ICAP, comparing similarities and differences is a critical strategy for promoting inference-making, leading to the C level of student engagement (Chi & Wylie, 2014; Chi et al., 2018). Other researchers (Rittle-Johnson & Star, 2009, 2011) have also emphasized the power of comparisons in mathematics learning. Therefore, to support PSTs' learning to teach mathematics, it is important to help them recognize when and how to ask "comparison" questions to elicit students' deep explanations. In conventional classrooms, however, Teacher Y rarely asked comparison questions, missing opportunities to promote deep learning—a common situation when lessons focus on computation rather than structure or relationships (Ding, 2021).

Additionally, across Chapters 3–6, Teacher X often began lessons with a real-world picture and invited students to pose questions. Problem-posing is a critical skill in mathematics learning (Cai, 1998; Cai & Rott, 2024; Cai et al., 2023; Silver & Cai, 1996), and I attended to these problem-posing questions when analyzing the contrasting cases, as they exemplify how students can achieve high levels of cognitive engagement according to ICAP (Chi et al., 2018). Therefore, I encourage PSTs to learn to invite students to pose questions, which fosters both creative and deep thinking as well as ownership of learning. Note that many PSTs in my methods course thought that one must follow the P-A-C-I order when teaching a lesson based on the ICAP model. However, problem-posing questions suggest that a teacher could immediately engage students at the A level without necessarily restricting their thinking to P.

7.3 Revisiting Contrasting Cases and ICAP

In this book, the TEPS approach is illustrated through contrasting cases. Three aspects should be emphasized. First, by "cases," I refer to the teaching of the worked example portion rather than a full lesson. This focus is intentional because well-taught worked examples can produce the "worked example effect" (Sweller & Cooper, 1985), helping students develop mental schemas for solving new problems. For this reason, the book emphasizes conceptual understanding over procedural fluency during this part of teaching, although both conceptual and procedural knowledge remain equally important (National Research Council, 2001).

Second, by "contrasting," I highlight the potential of comparison in supporting PSTs' learning to teach mathematics. Contrasting cases provide opportunities for comparisons, a process that demands higher order thinking skills. According to the ICAP framework (Chi et al., 2018), comparisons are critical for facilitating learners' high levels of cognitive engagement. However, researchers (Loibl et al., 2020) have found that contrasting cases without explicit guidance may not impact students' learning. Chin et al. (2016) also observed that when students are given contrasting cases, they often focus on surface features unless explicitly prompted to identify underlying principles. Therefore, I encourage math methods instructors to draw PSTs' attention to critical instructional features, as detailed in the "three-step" questions in earlier chapters. For example, PSTs can be asked to compare Teachers X and Y's lessons in terms of representation and questioning strategies, why these differ, and the potential impact on student learning.

Third, the analysis of contrasting cases is supported by ICAP. To familiarize PSTs with ICAP, I introduced it at the very beginning of the semester, asking them to read and respond to Chi and Boucher (2023) through professional writing. This understanding was then applied in analyzing the contrasting cases. However, since PSTs are future teachers rather than researchers, I only expected them to analyze Teacher X and Y's lessons holistically—considering the overall ICAP level—rather than with research-level rigor. I believe the ICAP framework serves as an effective tool for PSTs to understand the differences between deep and shallow learning in mathematics classrooms.

Thoughtful discussions of contrasting cases can contribute to PSTs' own lesson planning. In my pilot work, many PSTs initially developed lesson plans similar to Teacher Y's. After analyzing the contrasting cases, PSTs could refer to Teacher X's lesson to consider how Teacher Y's lesson could be enhanced and how their own plans could be improved. Therefore, in this book, I encourage PSTs to conduct preparatory work (Step 1 in earlier chapters) before reading and analyzing the contrasting cases (Step 2 in earlier chapters), and then reflecting on the cases, including comparing their initial plans with the contrasting cases (Step 3 in earlier chapters). These three steps have been piloted in my methods course through course preparation, in-class discussion, and written reports (either in narrative or tabular form). These processes appear to prompt PSTs to reflect deeply on how to incorporate the TEPS approach into their future planning and teaching.

7.4 Looking Back and Moving Forward

The TEPS approach in this book is illustrated through the teaching of only two early algebra concepts: Inverse relations and the basic properties of operations, which do not cover the full range of elementary mathematics topics. Nevertheless, I anticipate that TEPS—teaching through example-based problem solving—can be extended to other topics. For instance, the essential aspects of TEPS, concreteness fading and deep questioning, are likely transferable to teaching a variety of mathematical concepts beyond early algebra.

Additionally, contrasting cases have proven to be a powerful model for supporting PSTs' learning. In my pilot work, PSTs' reflections indicated that comparing the lessons of Teachers X and Y helped them articulate how TEPS, as opposed to conventional approaches, promoted students' deep learning. Moreover, the ICAP framework proved to be a feasible and valuable tool that PSTs could use to justify the contrasting cases and thus learn how to plan and teach of elementary mathematics.

Future work with PSTs could continue using the contrasting cases format alongside the ICAP framework and expand to other critical elementary concepts. With the growing availability of AI, this technology could be integrated into the development of contrasting cases. For example, AI tools (e.g., ChatGPT) can generate contrasting lesson plans (good and poor) that PSTs can be guided to compare, focusing on representations and deep questions. Some TEPS features may be absent even from the "good" versions, but these still provide opportunities for PSTs to critically evaluate the materials. Comparing them with plans produced by expert teachers can further promote PSTs' critical thinking skills. All of these analyses can be framed and justified using the ICAP framework. AI could also be used to transform written cases into more interactive formats. In our pilot work, we converted four pairs of written cases into VR-supported videos to better support PSTs' learning. The YouTube links shared in this book (see footnote in Chapter 3) showcase this trial work. Although the videos are not of the highest quality, we hope they serve as a useful resource for methods course instructors. In our pilot work, most PSTs preferred reading the written contrasting cases alongside the VR videos. We also hope that our work will inspire further innovations in supporting PSTs' lesson planning and and in promoting effective mathematics teaching through the integration of AI.

References

Association of Mathematics Teacher Educators. (2017). *Standards for preparing teachers of mathematics*. https://amte.net/standards

Cai, J. (1998). An investigation of U.S. and Chinese students' mathematical problem posing and problem solving. *Mathematics Education Research Journal*, *10*(1), 3–15. https://doi.org/10.1007/BF03217121

Cai, J., Ran, H., Hwang, S., Ma, Y., Han, J., & Muirhead, F. (2023). Impact of prompts on students' mathematical problem posing. *The Journal of Mathematical Behavior*, *72*. https://doi.org/10.1016/j.jmathb.2023.101087

Cai, J., & Rott, B. (2024). On understanding mathematical problem-posing processes. *ZDM – Mathematics Education*, *56*(1), 61–71. https://doi.org/10.1007/s11858-023-01536-w

Carpenter, T. P., Fennema, E., Franke, M. L., Levi, L., & Empson, S. B. (2015). *Children's mathematics, second edition: Cognitively guided instruction* (first ed.). 1999 Heinemann edition.

Carpenter, T. P., Franke, M. L., & Levi, L. (2003). *Thinking mathematically: Integrating arithmetic & algebra in elementary school*. Heinemann.

Chi, M. T., Adams, J., Bogusch, E. B., Bruchok, C., Kang, S., Lancaster, M., & Yaghmourian, D. L. (2018). Translating the ICAP theory of cognitive engagement into practice. *Cognitive Science*, *42*(6), 1777–1832. https://doi.org/10.1111/cogs.12626

Chi, M. T., & Boucher, N. S. (2023). Applying the ICAP framework to improve classroom learning. In C. E. Overson et al. (Eds.). *In their own words: What scholars want you to know about why and how to apply the science of learning in your academic setting* (pp. 94–110). American Psychological Association.

Chi, M. T., & Wylie, R. (2014). The ICAP framework: Linking cognitive engagement to active learning outcomes. *Educational Psychologist*, *49*(4), 219–243. https://doi.org/10.1080/00461520.2014.965823

Chin, D. B., Chi, M., & Schwartz, D. L. (2016). A comparison of two methods of active learning in physics: Inventing a general solution versus compare and contrast. *Instructional Science*, *44*(2), 177–195. https://doi.org/10.1007/s11251-016-9374-0

Ching, B. H. H., & Wu, X. (2019). Concreteness fading fosters children's understanding of the inversion concept in addition and subtraction. *Learning and Instruction*, *61*, 148–159. https://doi.org/10.1016/j.learninstruc.2018.10.006

Ding, M. (2021). *Teaching early algebra through example-based problem solving: Insights from Chinese and US elementary classrooms*. Routledge.
Ding, M., Byrnes, J., & Ke, X. (2023a). Rethinking the role of "quality instruction" in predicting algebraic learning within an opportunity-propensity framework: An exploratory cross-cultural study. *Journal of Educational Psychology*, *115*(2), 241–266. https://doi.org/10.1037/edu0000767
Ding, M., & Carlson, M. A. (2013). Elementary teachers' learning to construct high quality mathematics lesson plans: A use of IES recommendations. *The Elementary School Journal*, *113*, 359–385. https://doi.org/10.1086/668505
Ding, M., Chen, W., & Hassler, R. (2019). Linear quantity models in US and Chinese elementary mathematics classrooms. *Mathematical Thinking and Learning*, *21*, 105–130. https://doi.org/10.1080/10986065.2019.1570834
Ding, M., Hassler, R., & Li, X. (2021). Cognitive instructional principles in elementary mathematics classrooms: A case of teaching inverse relations. *International Journal of Mathematical Education in Science and Technology*, 52(8), 1195–1224. https://doi.org/10.1080/0020739X.2020.1749319
Ding, M., Li, X., Manfredonia, M., & Luo, W. (2023b). US and Chinese elementary teachers' noticing of cross-cultural mathematics videos. *Journal of Mathematics Teacher Education*, *26*, 211–239. https://doi.org/10.1007/s10857-021-09526-z
Fosnot, C. T., & Jacob, B. (2010). *Young mathematicians at work: Constructing algebra*. Heinemann.
Fyfe, E., McNeil, R., Son, N. M., & Goldstone, J. Y. (2014). Concreteness fading in mathematics and science instruction: A systematic review. *Educational Psychology Review*, *26*, 9–15. https://doi.org/10.1007/s10648-014-9249-3
Fyfe, E. R., McNeil, N. M., & Borjas, S. (2015). Benefits of "concreteness fading" for children's mathematics understanding. *Learning and Instruction*, *35*, 104–120. https://doi.org/10.1016/j.learninstruc.2014.10.004
Fyfe, E. R., & Nathan, M. J. (2019). Making "concreteness fading" more concrete as a theory of instruction for promoting transfer. *Educational Review*, *71*(4), 403–422.
Loibl, K., Tillema, M., Rummel, N., & van Gog, T. (2020). The effect of contrasting cases during problem solving prior to and after instruction. *Instructional Science*, *48*, 115–136. https://doi.org/10.1007/s11251-020-09504-7
Ma, L. (1999). *Knowing and teaching elementary mathematics: Teachers' understanding of fundamental mathematics in China and the United States*. Routledge.
Morris, A. K., Hiebert, J., & Spitzer, S. M. (2009). Mathematical knowledge for teaching in planning and evaluating instruction: What can preservice teachers learn? *Journal for Research in Mathematics Education*, *40*(5), 491–529. https://www.jstor.org/stable/40539354
National Research Council. (2001). *Adding it up: Helping children learn mathematics*. National Academy Press. https://www.nap.edu/catalog/9822/adding-it-up-helping-children-learn-mathematics
Ottmar, E., & Landy, D. (2017). Concreteness fading of algebraic instruction: Effects on learning. *Journal of the Learning Sciences*, *26*(1), 51–78. https://doi.org/10.1080/10508406.2016.1250212
Rittle-Johnson, B., & Star, J. R. (2009). Compared with what? The effects of different comparisons on conceptual knowledge and procedural flexibility for equation solving. *Journal of Educational Psychology*, *101*(3), 529. https://doi.org/10.1037/a0014224
Rittle-Johnson, B., & Star, J. R. (2011). The power of comparison in learning and instruction: Learning outcomes supported by different types of comparisons. In J. P. Mestre & B. H. Ross (Eds.), *The psychology of learning and motivation: Cognition in education* (pp. 199–225). Elsevier Academic Press. https://doi.org/10.1016/B978-0-12-387691-1.00007-7
Russell, S., Schifter, D., & Bastable, V. (2011). *Connecting arithmetic to algebra: Strategies for building algebraic thinking in the elementary grades*. Heinemann.
Schifter, D., Bastable, V., & Russell, S. J. (2016a). *Number and operations, part 1: Building a system of tens*. Casebook. *National Council of Teachers of Mathematics*.
Schifter, D., Bastable, V., & Russell, S. J. (2016b). *Number and operations, part 2: Making meaning for operations casebook. National Council of Teachers of Mathematics*.
Silver, E. A., & Cai, J. (1996). An analysis of arithmetic problem posing by middle school students. *Journal for Research in Mathematics Education*, *27*(5), 521–539. https://doi.org/10.5951/jresematheduc.27.5.0521
Sweller, J., & Cooper, G. A. (1985). The use of worked examples as a substitute for problem solving in learning algebra. *Cognition and Instruction*, *2*(1), 59–89. https://doi.org/10.1207/s1532690xci0201_3
Witzel, B. S. (2005). Using CRA to teach algebra to students with math difficulties in inclusive settings. *Learning Disabilities: A Contemporary Journal*, *3*(2), 49–60.

INDEX

Note: Page numbers in **bold** and *italics* refer tables and figures, respectively; page numbers followed by 'n' refer to notes.

addition, teaching of: associative property of 90–103; commutative property of 79–90; conventional approach 83–85, 95–98; TEPS approach 80–83, 91–94

additive inverses, teaching of: comparison model 23–36; conventional approach 10, 16–19, 28–31; part-whole model 11–23; TEPS approach 10, 12–15, 25–28

array model of teaching multiplicative inverses 50–63; analysis of cases 58–63; contrasting cases 51; conventional approach 55–58; deep questions 59, **60**, 61; ICAP, linking to **60**, 62; insights **60**, 62–63; inverse word problems **51**; representation uses 59, **60**; TEPS approach 51–55

Association of Mathematics Teacher Educators 1, 144

associative property of addition, teaching of 90–103; analysis 99–103; contrasting cases 91; conventional approach 95–98; deep questions 99, **100**, 101–102; ICAP, linking to 102; insights 103; representation uses 99; TEPS approach 91–94

associative property of multiplication, teaching of 116–129; analysis of case 124–129; contrasting cases 117; conventional approach 122–124; deep questions **125**, 126–127; ICAP, linking to **125**, 127–128; insights 128–129; representation uses **125**, 126; TEPS approach 117–121

basic properties of operations 3; addition (*see* addition, teaching of); multiplication (*see* multiplication, teaching of)

Boucher, N. S. 3, 8, 36, 147

Chi, M. T. H. 3, 7, 8, 36, 147

cognitive engagement 2, 7–8, 10, 22–23, 34–36, 49, 62, 76, 78, 79, 89, 102, 104, 105, 115–116, 127–128, 140–142, 144–147

College of Education and Human Development (CEHD) 4*n*1

commutative property of addition, teaching of 79–90; analysis of case 86–90; contrasting cases 80; conventional approach 83–85; deep questions 86, **87**, 88; ICAP, linking to **87**, 89; insights 89–90; representation uses 86, **87**; TEPS approach 80–83

commutative property of multiplication, teaching of 105–116; analysis of case 113–116; contrasting cases 106; conventional approach 110–112; deep questioning 113–115, **114**; ICAP, linking to 115–116; insights 116; representation uses 113, **114**; TEPS approach 106–110

comparison model of teaching additive inverse 23–36; analysis of cases 32–36; contrasting cases 24; conventional approach 24, 28–31; deep questions **33**, 34–35; ICAP, linking to 32, **33**, 35; insights for planning and teaching 35–36; inverse word problems **24**; representation uses 32, **33**; TEPS approach 25–28

comparison model of teaching multiplicative inverses 63–77; analysis of case 73–77; contrasting cases 63–64; conventional approach 69–72; deep questions 73, **74**, 75–76; ICAP, linking to **75**, 76; insights **77**; representation uses 73, **74**; TEPS approach 64–69

comparison questions 7, **20**, 21–22, **33**, 34–35, **47**, 48–49, **60**, 61, **74**, 76, **87**, 88, 89, 102, 115, 127, 139–140, 146

concreteness fading 7, 10, 19, 22, 32, 35, 36, 38, 46, 50, 59, 62, 73, 78, 86, 89, 99, 103, 104, 113, 116, 126, 127, 137, 142, 144, 145, 147
contrasting cases 2–3, 147; addition teaching, associative property 91; addition teaching, commutative property of 80; additive inverse teaching, comparison model 24; additive inverse teaching, part-whole model 11; multiplication teaching, associative property 117; multiplication teaching, commutative property 106; multiplication teaching, distributive property 129–130; multiplicative inverses teaching, array model 51; multiplicative inverses teaching, comparison model 63–64; multiplicative inverses teaching, equal-groups model 39
Contrasting Cases Analyses (CCA) 3

deep questioning 7, 146; addition teaching, associative property 99, **100**, 101–102; addition teaching, commutative property 86, **87**, 88; additive inverse teaching, comparison model **33**, 34–35; additive inverse teaching, part-whole model of 19, **20**, 21–22; multiplication teaching, associative property **125**, 126–127; multiplication teaching, commutative property of 113–115, **114**; multiplication teaching, distributive property of 137, **138**, 139–140; multiplicative inverses teaching, array model 59, **60**, 61; multiplicative inverses teaching, comparison model 73, **74**, 75–76; multiplicative inverses teaching, equal-groups model 46, **47**, 48–49
Ding, M. 2
distributive property (DP) of multiplication over addition 129–141, 142*n*1; analysis of case 137–141; contrasting cases 129–130; conventional approach 134–136; deep questioning 137, **138**, 139–140; ICAP, linking to **139**, 140–141; insights 141; representation uses 137, **138**; TEPS approach 130–133

early algebra 3, 8, 144, 145, 147
equal-groups model of teaching multiplicative inverses 38–50; analysis of cases 46–50; contrasting cases 39; conventional approach 43–45; deep questions 46, **47**, 48–49; ICAP, linking to 49; insights 49–50; inverse word problems **39**; representation uses 46, **47**; TEPS approach 40–43

inference-making 22, 23, 34, 48, 62, 76, 86, 88, 89, 101, 103, 104, 115–116, 127–128, 140–142, 144, 146
Interactive, Constructive, Active, and Passive (ICAP) framework 2, 6, 145; addition, associative property 102; addition, commutative property **87**, 89; additive inverse teaching, comparison model 32, **33**, 35; additive inverse teaching, part-whole model **20**, 22–23; multiplication, associative property of **125**, 127–128; multiplication, commutative property of 115–116; multiplication, distributive property of **139**, 140–141; multiplicative inverses teaching, array model **60**, 62; multiplicative inverses teaching, comparison model **75**, 76; multiplicative inverses teaching, equal-groups model 49; student learning levels 7–8
inverse relations 3, 10, 11, 19, 24, 32, 39, 46, 49, 61, 63, 73, 76, 146, 147

meaning-making questions **20**, 21, 22, **33**, 34, **47**, 48, 59, **60**, **74**, 75–76, **87**, 88, 101–102, 114–115, 126–127, 139, 146
multiplication, teaching of: associative property of 116–129; commutative property of 105–116; conventional approach 110–112, 122–124, 134–136; distributive property of 129–141; TEPS approach 106–110, 117–121, 130–133
multiplicative inverses, teaching of: array model 50–63; comparison model 63–77; conventional approach 38, 43–45, 55–58, 69–72; equal-groups model 38–50; TEPS approach 38, 40–43, 51–55, 64–69

part-whole model of teaching additive inverse 11–23; analysis of cases 19–23; contrasting cases 11; conventional approach 16–19; deep questions 19, **20**, 21–22; ICAP, linking to **20**, 22–23; insights 23; part-part-whole model 19, *19*; part-whole inverse word problems **11**; representation uses 12, *12–18*, 19, **20**, 21; TEPS approach 12–15
preservice teacher (PST) 1–3, 8, 10, 23, 35–36, 49–50, 62–63, 77, 89–90, 103–104, 116, 128–129, 141, 144, 146–148
problem-posing questions 7, **20**, 21, **33**, 34, **47**, 48, 59, **60**, 73, **74**, 86, **87**, 99, **100**, 101, 146

tape diagrams 10, 15, 26, 27, 32, 34, 35, 38, *40–42*, 41–42, 46, 48, 49, *66–68*, 73, 75, 76, 78
Teaching Through Example-Based Problem Solving (TEPS) 2, 144; addition, associative property 91–94; addition, commutative property 80–83; additive inverse teaching, comparison model 25–28; additive inverse teaching, part-whole model 12–15; deep questions 7, 146; multiplication, associative property of 117–121; multiplication, commutative property of 106–110; multiplication, distributive property of 130–133; multiplicative inverses teaching, array model 51–55; multiplicative inverses teaching, comparison model 64–69; multiplicative inverses teaching, equal-groups model 40–43; representation uses 6–7, 145–146

VR-supported videos 4, 4*n*1, 148